Chemistry Tests for First Examinations

Also from Stanley Thornes & Hulton:

Atkinson:	Modern Organic Chemistry (3rd Edition)
Barlex:	Chemistry Cornerstones – A Third Year Chemistry Reader
Gilmore:	A Modern Approach to Comprehensive Chemistry (3rd Edn)
Ramsden:	A First Chemistry Course
Ramsden:	A-Level Chemistry
Ramsden:	Calculations for A-Level Chemistry
Ramsden:	Calculations for GCSE Chemistry

Chemistry Tests for First Examinations

A. PORTER, BSc
Assistant Chemistry Master at Batley High School for Boys

T. WOOD, BSc
Head of Chemistry and Head of Sixth Form at Batley High School for Boys

Stanley Thornes (Publishers) Ltd

First published in 1987 by:
Stanley Thornes (Publishers) Ltd
Old Station Drive
Leckhampton
CHELTENHAM GL53 0DN
England

Reprinted 1988

British Library Cataloguing in Publication Data

Porter, A.
Chemistry tests for first examinations.
1. Chemistry — Examinations, questions, etc.
I. Title II. Wood, T.
540'.76 QD42

ISBN 0-85950-666-5

Typeset by Tech-Set, Gateshead, Tyne and Wear in 10/12 Palatino
Printed and bound in Great Britain at The Bath Press, Avon

Contents

Tests

Reaction Schemes

Preface

The major part of this book consists of thirty tests suitable for use with the majority of pupils taking a first external examination in chemistry. Each test is designed to be taken during a single period of approximately half an hour, and is worth thirty marks. The tests start with easy questions and gradually become more demanding in terms of both reading ability and answers required. They have been extensively tested by pupils at Batley High School for Boys, to whom the authors extend their thanks. Questions vary in types within the tests, and no attempt has been made to produce a pattern. The tests should cover most of the content of all new syllabuses in use, and proposed, for pupils up to the age of sixteen.

The second section consists of twenty-five reaction scheme tests, each worth ten marks. These are much shorter than the other tests, but we have found them very useful indeed as revision aids. Pupils also enjoy them if they are approached as detective stories in which clues are laid to be solved.

A booklet of answers is also available from the publishers.

Batlium, the wonder element used in many tests, is dedicated to the alchemists of old. They searched in vain for a universal solvent, without realising that if they found it, they would not be able to find anything to keep it in.

A. PORTER
T. WOOD
Batley, West Yorkshire,
1987

Acknowledgements

The copper(II) sulphate crystal on the cover was grown by fourteen-year-old pupils at Batley High School for Boys, under the supervision of Andrew Porter. The photograph was taken by Trevor Wood.

TESTS

TEST 1
Elements, Compounds, Mixtures and Changes in State

1 **Matching pairs question**. For questions (a) to (c) there are five alternative answers, **A** to **E**. Each letter may be used *once, more than once* or *not at all*. Show your choice by writing the letter opposite the number on your answer sheet.

A **oxygen**

B **air**

C **carbon dioxide**

D **water vapour**

E **nitrogen**

a) Which *one* gas is a mixture? (1)

b) Which *two* gases are compounds? (2)

c) Which *two* gases are elements? (2)

2 Copy out and complete this paragraph.

Carbon dioxide is a ____________ at room temperature and pressure. When cooled down it becomes a ____________ known as dry ice. This change in state is known as ____________ because the carbon dioxide became dry ice without going through the ____________ state. (4)

3 What words take the place of the letters **A** to **D** on the diagram below?

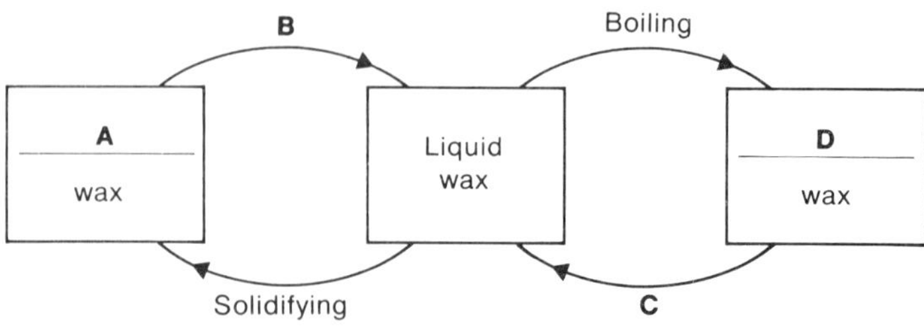

(4)

4 The diagram below shows hydrogen being burned in air and the product being collected.

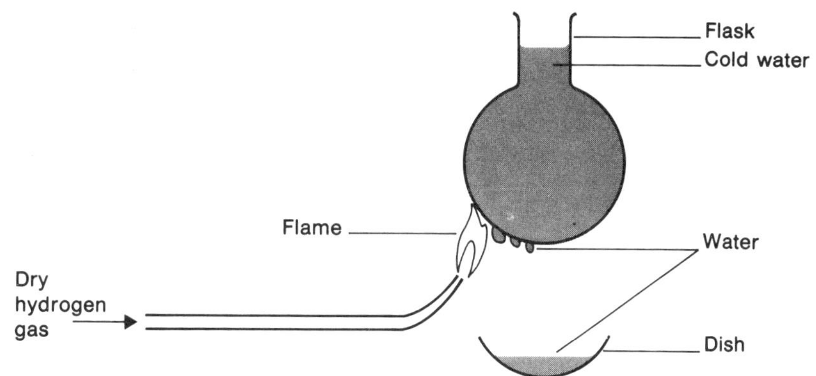

a) Has the hydrogen undergone a physical or chemical change? (1)

b) Give a reason for your answer to (a). (1)

c) If 10 g of water were collected and evaporated, how many grams of steam would be produced?

A **5 g** **B** **9 g** **C** **10 g** **D** **11 g** **E** **20 g** (1)

d) In evaporating, has the water undergone a physical or chemical change? (1)

e) Give a reason for your answer to (d). (1)

5 Give the common name for the simplest compound formed between the following elements.

a) hydrogen and oxygen (1)

b) nitrogen and hydrogen (1)

6 **Matching pairs question**. For questions (a) to (e) there are five alternative answers, **A** to **E**. Each letter may be used *once, more than once* or *not at all*. Show your choice by writing the letter opposite the number on your answer sheet.

A **alloy** **B** **compound** **C** **element**

D **mixture** **E** **solution**

Which word from the list above *best* describes

a) common salt (1)

b) sea water (1)

c) brass (1)

d) chlorine (1)

e) rock salt? (1)

7 A new substance, named batlium, has been discovered. It has a melting point of −20 °C and it boils at 85 °C.

a) Is batlium a solid, liquid or gas at room temperature? (1)

b) If a mixture of batlium and water was distilled, which substance would be collected first? (1)

c) What effect would a decrease in atmospheric pressure have on the boiling point of batlium? (1)

d) What *two* things would happen to its melting point if the batlium was impure? (2)

TEST 2
Separating Techniques

1 **Matching pairs question.** For questions (a) to (e) there are five alternative answers, **A** to **E**. Each letter may be used *once, more than once* or *not at all.* Show your choice by writing the letter opposite the number on your answer sheet.

A **filtrate**

B **residue**

C **distillate**

D **immiscible**

E **miscible**

Which word best fits the following meanings?

a) the solid left on a filter paper after filtration (1)

b) alcohol and water being two liquids which mix in all proportions (1)

c) the anhydrous copper(II) sulphate left after a solution of copper(II) sulphate has been heated to dryness (1)

d) the liquid collected after filtration (1)

e) tetrachloromethane and water forming two layers (1)

2 How would you separate

a) oil from a mixture of oil and water (2)

b) iron filings from a mixture of iron filings and sand (2)

c) sand from a mixture of sand and water? (2)

3 A solution of sodium chloride can be separated into solid sodium chloride and pure water by the process called distillation. The diagram below shows the apparatus needed.

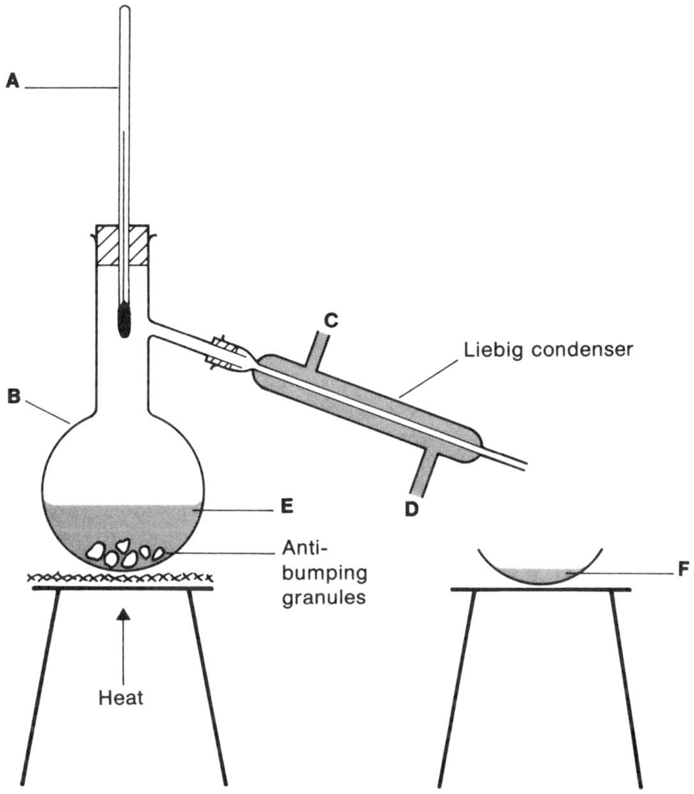

a) Why are anti-bumping granules added? (1)

b) Name the pieces of apparatus **A** and **B**. (2)

c) Does the cooling water enter the apparatus at **C** or **D**? (1)

d) Give a reason for your choice in (c). (1)

e) Which letter is pointing to the pure water? (1)

f) Why would it be easier to separate a mixture of calcium carbonate and water? (1)

4 A blackmail letter was suspected of being written by one of three pens belonging to Miss X, Mrs Y and Mr Z. A sample of ink from the letter was placed alongside a sample from the three pens on a sheet of filter paper. This was dipped into a tank containing a volatile solvent.

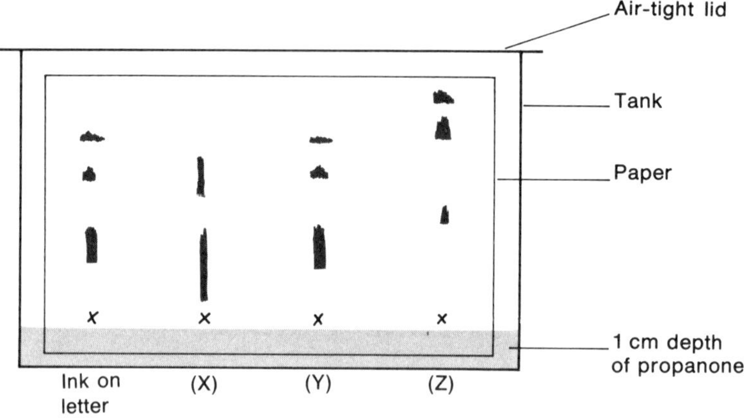

a) What name is given to this process? (1)

b) Why is an airtight lid used? (1)

c) Why would the test be spoilt if 2 cm of propanone was used? (1)

d) Who is the most likely blackmailer? (1)

e) Give a reason for your answer. (1)

5 a) Draw a complete, labelled and neat diagram showing the apparatus needed to separate a mixture of sodium chloride and ammonium chloride. (5)

b) What happens to the ammonium chloride in this experiment? (1)

c) What happens to the sodium chloride in this experiment? (1)

TEST 3
Acids, Bases and Salts

1 Acids are chemicals which are often found in nature and have many uses in daily life. For the table on the opposite page, say what should go in place of the letters **A**, **B**, **C** and **D**.

Name of acid	*Use or occurrence of acid*
A	Car battery acid
Ethanoic acid	**B**
Citric acid	**C**
D	In all fizzy drinks

(4)

2 What name is given to the group of chemicals used to identify acids and alkalis by a colour change? (1)

3 Copy out and complete this passage.

Bases are defined as chemicals which react with acids, i.e. accept ____________ ions, to form a ____________ and ____________ only. A base which dissolves in water is called an ____________ . (4)

4 Below is a diagram of an experiment involving sodium hydroxide and hydrochloric acid (both as dilute solutions).

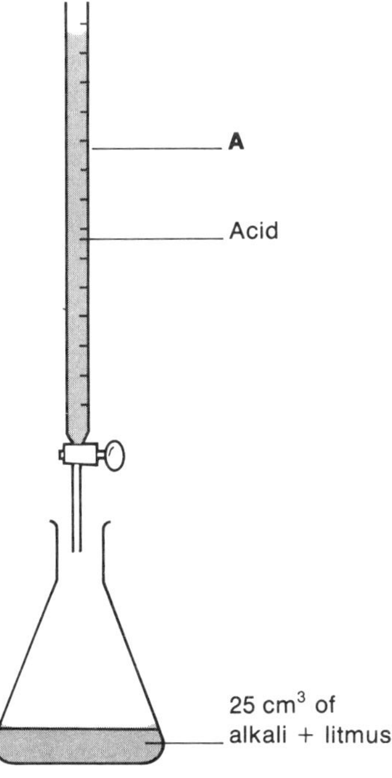

a) What colour is litmus in alkali? (1)

b) Name the piece of apparatus labelled **A**. (1)

c) Name the type of reaction taking place. (1)

d) It took 28.00 cm^3 of acid to react exactly with the alkali. Which solution is more concentrated? (1)

e) This equation shows the reaction between sulphuric acid and sodium hydroxide.

$$H_2SO_4(aq) + NaOH(aq) \rightarrow Na_2SO_4(aq) + H_2O(l)$$

Write out this equation and balance it. (1)

f) If the sulphuric acid is in the same concentration as was the hydrochloric acid in the experiment above, what volume of sulphuric acid would be needed to react exactly with 25 cm^3 of the same sodium hydroxide solution? (1)

g) Give a reason for your answer to the above question. (2)

5 What salt is formed when calcium hydroxide reacts with nitric acid? (1)

6 What are the products of the following reaction?

Zinc oxide + nitric acid → (1)

7 In the list below are five versions, **A** to **E**, of the three stages in the preparation of potassium sulphate from potassium hydroxide solution and sulphuric acid. Which is correct?

	1st stage	*2nd stage*	*3rd stage*
A	Evaporation	Neutralisation	Crystallisation
B	Evaporation	Crystallisation	Neutralisation
C	Neutralisation	Evaporation	Crystallisation
D	Crystallisation	Evaporation	Neutralisation
E	Neutralisation	Crystallisation	Evaporation

(1)

8 Silver iodide can be prepared by reacting silver nitrate solution with sodium iodide solution. Silver iodide is insoluble in water.

a) What would you observe in such a reaction? (1)

b) How would you obtain a dry sample of silver iodide? (1)

c) Silver nitrate is itself a salt. Which acid must be used to form this silver salt? (1)

9 Water molecules partially ionise to form hydrogen ions ($H^+(aq)$).

a) Write an ionic equation for the formation of the ions. (1)

b) Why doesn't water show acidic properties? (2)

10 The two equations below compare the dissolving of hydrogen chloride and ethanoic acid in water.

$$HCl(g) \rightarrow H^+(aq) + Cl^-(aq)$$

$$CH_3COOH(l) \rightleftharpoons H^+(aq) + CH_3COO^-(aq)$$

a) Why is hydrochloric acid a strong acid? (1)

b) Why is ethanoic acid a weak acid? (1)

c) What would be a likely pH value for each of the acids? (2)

TEST 4
Atomic Structure and Bonding

1 The atom is made up of a central part, which is dense, surrounded by a less dense part.

a) What particle or particles would you expect to find in the central part? (2)

b) What is the name given to this central part? (1)

c) What particle or particles would you expect to find in the less dense part? (1)

2 Copy and complete the following sentence.

Atoms are electrically neutral, and so they must contain equal numbers of ____________ and ____________ . (2)

3 Which atom contains no neutrons? (1)

4 Which one of the following numbers of electrons in the outer energy level is possible for the newly discovered wonder element batlium, which is found to be in Group 7?

A 4 B 5 C 6 D 7 E 8 (1)

5 The number of electrons involved in the triple bond in nitrogen (atomic number = 7) is

A 2 B 3 C 4 D 5 E 6. (1)

6 What is the name given to the elements in Group 0? (1)

7 **Matching pairs question.** For questions (a) to (c) each question has four alternative answers, **A** to **D**. Each letter may be used *once, more than once* or *not at all*. Show your choice by writing the correct letter at the side of the question number on your answer sheet. From the list

A sodium chloride

B iodine

C carbon dioxide

D chlorine

choose the chemical which

a) has the highest melting point (1)

b) is a covalent compound which sublimes (1)

c) is a diatomic covalent gas at room temperature and pressure. (1)

8 Explain how a covalent bond is formed using, as an example, *one* of the following (atomic numbers: H = 1, O = 8, Cl = 17).

hydrogen gas **water** **hydrogen chloride** (2)

9 Copy and complete the following sentence.

An ion is an atom or ____________ which has ____________ or ____________ one or ____________ electrons. (2)

10 What is the name given to atoms of an element which contain equal numbers of protons and electrons but have different numbers of neutrons? (1)

11 Rubidium is a Group 1 metal which is more reactive than sodium.

Group 1	
Lithium	Li
Sodium	Na
Potassium	K
Rubidium	Rb
Caesium	Cs
Francium	Fr

When it reacts with oxygen, it forms a solid **A**. It reacts in water to form a gas **B**, and a strongly alkaline solution **C**. Use this information to help answer the following questions.

a) Give the name of the group of metals of which rubidium is a member. (1)

b) How many electrons are there in the highest energy level in rubidium? (1)

c) What would be the expected valency of rubidium? (1)

d) Give the name of the solid **A**. (1)

e) Give the formula of solid **A**. (1)

f) Give the name of gas **B**. (1)

g) Give the formula of gas **B**. (1)

h) Give the name of solution **C**. (1)

i) Give the formula of solution **C**. (1)

j) What type of bonding is present in solid **A**? (1)

k) What can you say about the melting point of solid **A**? (1)

l) What type of bonding is present in gas **B**? (1)

TEST 5
The Periodic Table

1

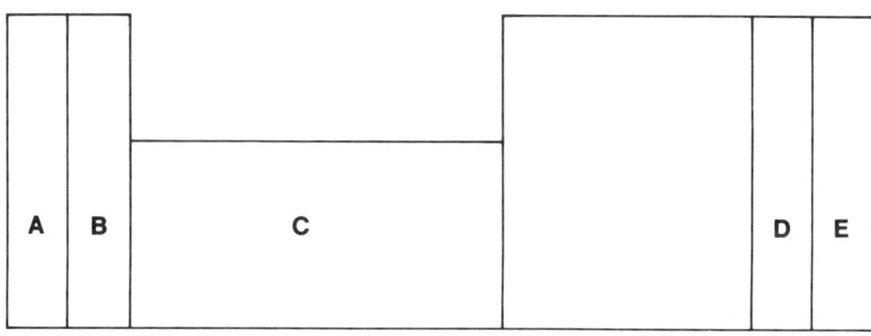

On this rough representation of the periodic table, which area corresponds to

a) transition metals (1)

b) noble gases (1)

c) Group 7 elements (1)

d) Group 2 elements? (1)

2 Here is an element taken from the periodic table.

27 Al 13

a) What is the name of this element? (1)

b) What is the atomic number of the element? (1)

c) How many electrons does an atom of the element have? (1)

d) What is the electron arrangement? (1)

e) How many neutrons does an atom of the element have? (1)

f) What group does the element belong to? (1)

3 Copy and complete the following equations. If there would be no reaction then simply write, *'no reaction'*.

a) $Cl_2(aq) + K^+Br^-(aq) \rightarrow$ (1)

b) $Cl_2(aq) + K^+Cl^-(aq) \rightarrow$ (1)

c) $Br_2(aq) + K^+I^-(aq) \rightarrow$ (1)

4

Element	*Sodium*	*Magnesium*	*Aluminium*	*Silicon*	*Phosphorus*	*Sulphur*
Formula of oxide	Na_2O	MgO	Al_2O_3	SiO_2	P_2O_5	SO_2

The table above shows a period across the periodic table.

a) Name *one* non-metal from this period. (1)

b) Name *one* metal from this period. (1)

c) Does the size of the atoms of the elements increase or decrease across the period (left to right)? (1)

d) Give a reason for your answer to (c). (1)

e) Name one oxide which will produce an acidic solution. (1)

f) Which oxide is amphoteric? (1)

5 **Matching pairs question**. For questions (a) to (d) there are five alternative answers, **A** to **E**. Each letter may be used *once, more than once* or *not at all.* Show your choice by writing the letter opposite the number on your answer sheet.

A alkali metals **B alkali earth metals** **C halogens**

D noble gases **E transition metals**

a) Which of these are elements with one electron in their outer shell? (1)

b) Which of these have characteristically coloured compounds? (1)

c) Which of these are very unreactive towards other elements? (1)

d) Which of these is a group in which the elements decrease in reactivity down the group? (1)

6 Transition metals are often described as ductile. What does this mean? (1)

7 Below is a table showing three elements found on another planet, together with their electronic structures.

Name of element	*Electronic structure*
Batlium	2.8.2
Woodium	2.8.8.2
Portium	2.8.18.8.2

a) Which group of the periodic table would we put these elements in? (1)

b) Which element has the largest atoms? (1)

c) Which element would be the most reactive? (1)

d) Give a reason for your answer to (c). (2)

e) What would be the electron arrangement for the element above batlium in the group? (1)

TEST 6
Air and Oxygen

1 Which one of the following is correct for the approximate percentage of nitrogen in the air?

A 20% B 40% C 60% D 80% E 100% (1)

2 Which one of the following metals can rust?

A aluminium B copper C iron D lead E tin (1)

3 What is the test for oxygen gas? (1)

4 Combustion is the process which happens when chemicals burn in air.

a) What gas, in the air, is needed to help burning? (1)

b) Name X and Y in the fire triangle shown below. (2)

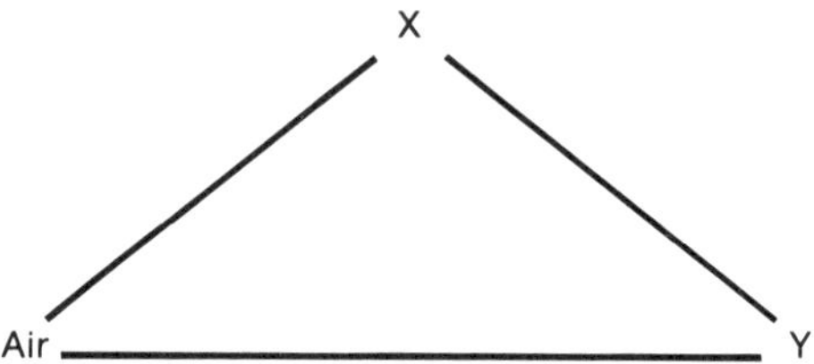

5 The diagram below represents an experiment to find the percentage composition of air.

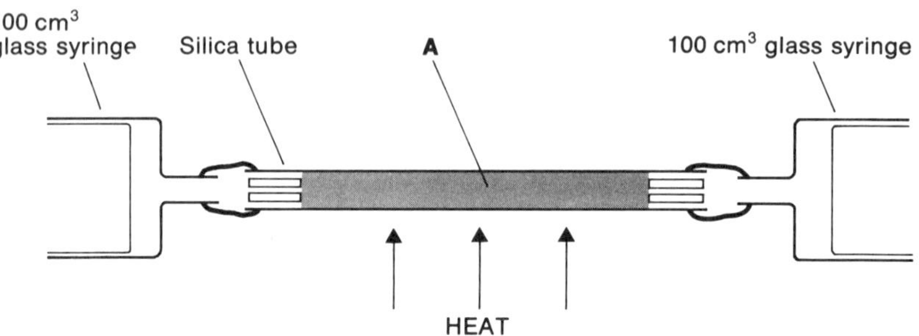

a) Name chemical **A**. (1)

b) What substance in the air is removed by this heated chemical? (1)

c) What can we *see* happening inside the silica tube which tells us that a chemical reaction is taking place? (1)

d) Why do we use a silica tube rather than an ordinary glass tube? (1)

6 Write out and complete the following.

Humans breathe in air. In fact, we need the ____________ in the air to live. We breathe out more carbon ____________ than we breathe in. The air we breathe out is also saturated with ____________ ____________. (4)

7 The percentage of oxygen in exhaled air is approximately

A 16% **B** 17% **C** 18% **D** 19% **E** 20%. (1)

8 With the aid of a sketch, explain briefly an experiment you could do in the laboratory, to show that air contains water vapour. (2)

9 Some new houses are built at the side of a main road about two miles from the centre of a city. Pollution monitoring equipment is set up in their gardens, and twice a day a rise in the level of pollution is recorded.

a) What gas is the main cause of the pollution? (1)

b) How is this gas produced? (1)

c) At what *times* of the day does the pollution increase? (2)

10 Study the data below.

Gas	*Melting point* (°C)	*Boiling point* (°C)
Nitrogen	−210	−196
Oxygen	−219	−183

In order to separate nitrogen from oxygen, the air is first liquefied.

a) During liquefaction, at what temperature will nitrogen first become a liquid? (1)

b) Which is easier to liquefy, oxygen or nitrogen? (1)

c) Explain your answer to part (b). (1)

d) Give one large-scale use of nitrogen. (1)

11 A spacecraft from Earth finds a planet in a distant galaxy where the air has the following composition.

Substance	*Oxygen*	*Nitrogen*	*Carbon dioxide*	*Argon*	*Neon*
%	17.00	63.00	0.05	6.95	13.00

The planet is the same size and density as Earth, and is as warm (15 °C average temperature).

a) Will the crew be able to live in this atmosphere? (1)

b) Explain your answer to (a) above. (1)

c) Assuming the above figures to be very accurate, what substance is missing from the air, which is present in Earth's atmosphere? (1)

d) Which two gases are present in the air on this planet in far larger amounts than on Earth? (2)

TEST 7
Water

1 Rain water, if free from pollutants, is normally weakly acidic. The chemical which is the cause of this acidity is

A **oxygen**

B **nitrogen**

C **carbon dioxide**

D **sulphur dioxide**

E **nitrogen oxide.** (1)

2 Which one of the following is an advantage of hard water?

A **It may help to produce sound teeth.**

B **It uses a lot of soap.**

C **It produces a scum with soaps.**

D **It may produce a white fur in hot water pipes which acts as an insulator.**

E **It increases the sales of detergents in local shops.** (1)

3 **Matching pairs question**. For questions (a) to (e) there are five alternative answers, **A** to **E**. Each letter may be used *once, more than once* or *not at all*. Show your choice by writing the letter opposite the number on your answer sheet. From the list **A** to **E** below

A **calcium carbonate**

B **calcium hydrogencarbonate**

C **ammonium nitrate**

D **calcium sulphate**

E **anhydrous copper(II) sulphate**

choose the substance which

a) is present in temporarily hard water (1)

b) is present in permanently hard water (1)

c) is a cause of pollution in rivers (1)

d) is the scale on kettles in some hard water areas (1)

e) may be used to test for the presence of water. (1)

4 Some water which had been boiled for half an hour was cooled to room temperature. Fish were then placed in it. The fish soon died. Why? (2)

5 a) Which non-metallic ion is present in water which is temporarily hard? (1)

b) How can temporarily hard water be most easily softened? (1)

c) Why is some water said to be permanently hard? (1)

d) Which chemical may be added to remove permanent hardness? (1)

6 Below is a representation of the water cycle.

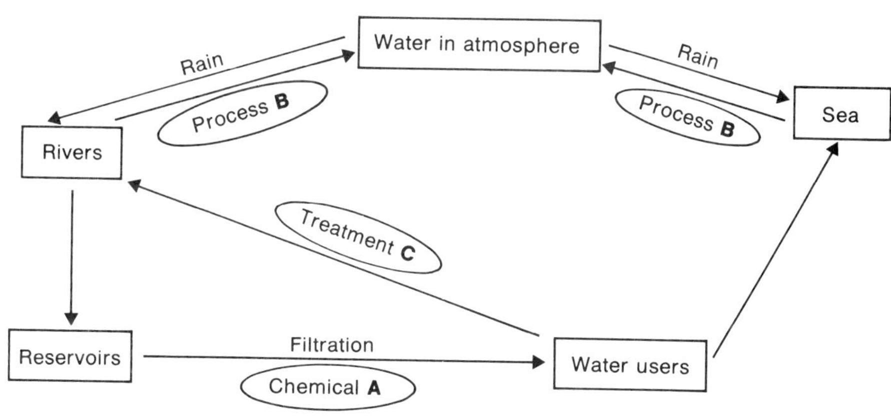

a) What is chemical **A**? (1)

b) What is process **B**? (1)

c) What is treatment **C**? (1)

7 Why are some water authorities so keen to add a fluoride to the water supply? (1)

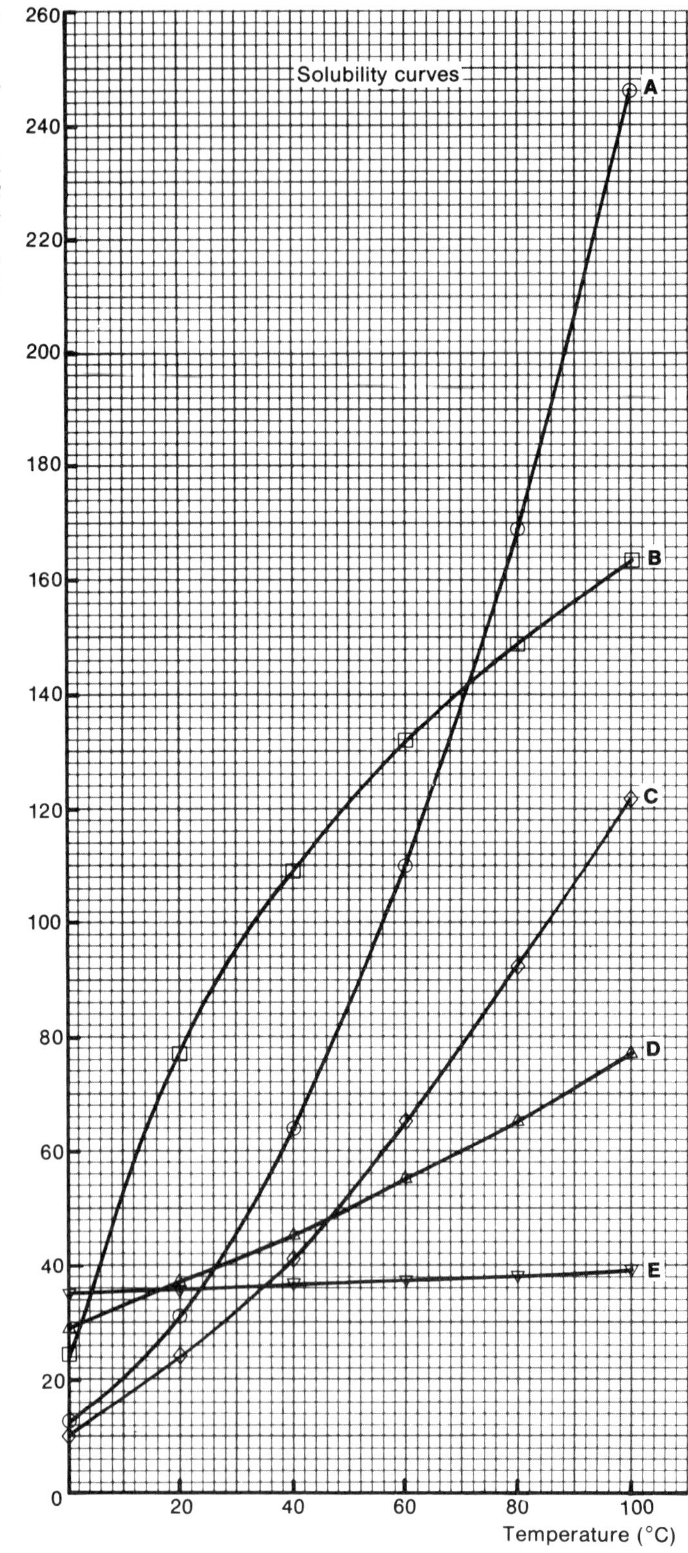
Solubility curves
Solubility (g per 100 g water)
260
240
220
200
180
160
140
120
100
80
60
40
20
0
20
40
60
80
100
Temperature (°C)
A
B
C
D
E

8 Give a balanced symbol equation for the reaction between hydrogen and oxygen. (2)

9 Why does tap water sometimes appear 'cloudy' when first run into a clean beaker? (1)

10 Study the following data. Graphs using the data for this question appear opposite.

Solubility in g per 100 g water

Temperature (°C)	*Chemicals*		
	Sodium chloride	Potassium nitrate	Ammonium chloride
0	35.7	13.3	29.4
20	36.0	31.3	37.2
40	36.6	63.9	45.0
60	37.3	110	55.0
80	38.4	169	65.0
100	39.8	246	77.3

a) Which of the graphs **A** to **E** opposite best represents the solubility curve of

i) sodium chloride (1)

ii) potassium nitrate (1)

iii) ammonium chloride? (1)

b) Estimate, to the nearest whole g per 100 g water, the solubility of

i) ammonium chloride at 50 °C (1)

ii) ammonium chloride at 70 °C. (1)

c) How much potassium nitrate will dissolve in 1000 g water at 20 °C? (1)

d) How much sodium chloride will dissolve in 200 g water at 20 °C? (1)

e) How much potassium nitrate will dissolve in 10 g water at 60 °C? (1)

11 Why do soapless detergents lather in hard water while soaps produce a scum? (2)

TEST 8
Hydrogen

1 What is the test for hydrogen gas? (1)

2 Which one of the following is a metal which will *not* produce hydrogen with dilute hydrochloric acid?

A **calcium** B **copper** C **iron** D **magnesium** E **zinc** (1)

3 Which one of the following produces hydrogen most rapidly in cold water?

A **calcium** B **lithium** C **magnesium** D **potassium** E **sodium** (1)

4 a) What is the *name* of the chemical which is made when hydrogen burns in air? (1)

b) Give a word equation for this reaction. (1)

c) Give a balanced symbol equation for this reaction. (2)

5 Below is a diagram for the reaction between steam and heated magnesium ribbon.

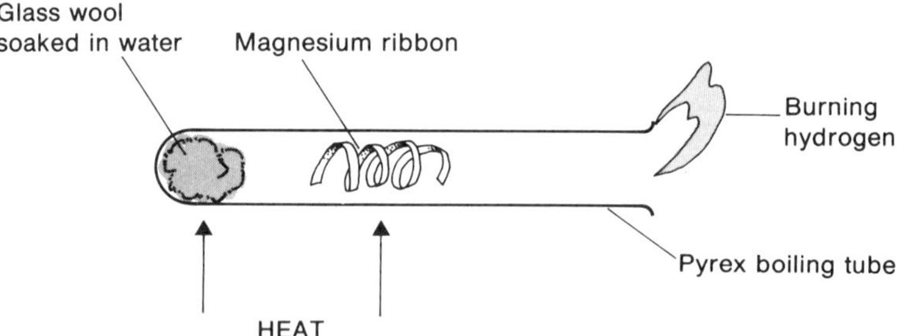

a) Give a word equation for this reaction. (1)

b) Give a balanced symbol equation for this reaction. (2)

c) Why do you not look directly at the apparatus during the reaction? (1)

d) Why is it necessary to use a pyrex boiling tube rather than a soda glass one? (1)

e) Why is the reaction of magnesium with steam faster than that with cold water? (1)

f) If the magnesium were replaced by zinc, would the reaction be faster or slower? (1)

g) Explain your answer to (f) above. (1)

6 In the laboratory hydrogen is often prepared using zinc granules and dilute hydrochloric acid.

a) Give a word equation for this preparation. (1)

b) Give a balanced symbol equation for this reaction. (2)

7 **Matching pairs question**. Questions (a) to (c) have five alternative answers, **A** to **E**. Each letter may be used *once, more than once* or *not at all*. Show your choice by writing the letter opposite the question number on your answer sheet.

Dry hydrogen gas may be used as a reducing agent, as shown in the diagram below.

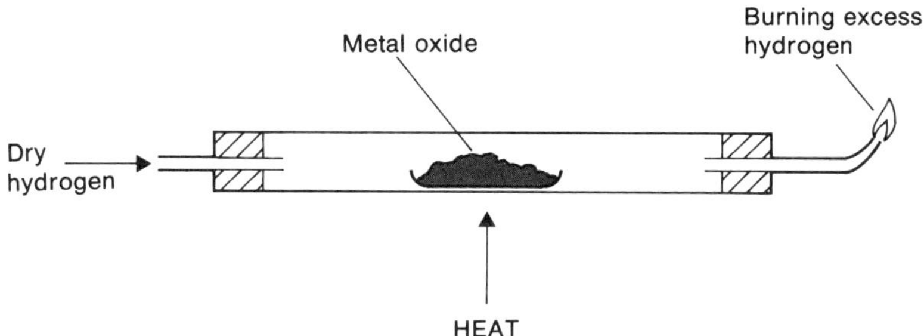

From the list **A** to **E** below

A **concentrated sulphuric acid**

B **copper(II) oxide**

C **magnesium oxide**

D **carbon dioxide**

E **silica gel**

choose the substance which

a) is the best for drying the hydrogen (1)

b) may be used as the metal oxide in the reaction (1)

c) is not a suitable metal oxide for the reaction. (1)

8 In the experiment above for question 7, explain why

a) it is necessary to allow plenty of hydrogen through the apparatus before lighting the excess gas (1)

b) after reaction, the metal is cooled in a stream of hydrogen. (1)

9 a) Give *one* advantage of the use of hydrogen as a fuel. (1)

b) Give *one* disadvantage of the use of hydrogen as a fuel. (1)

10 **Matching pairs question**. Questions (a) to (e) have nine alternative answers, **A** to **I**. Each letter may be used *once, more than once* or *not at all.* Show your choice by writing the letter opposite the question number on your answer sheet. From the list **A** to **I** below

A **boiling point**

B **coloured**

C **harden**

D **hydrogenation**

E **melting point**

F **saturated**

G **soften**

H **sweeter**

I **unsaturated**

choose the letter which best answers the following questions about the use of hydrogen in the manufacture of margarine from edible oils.

a) What word may be used to describe this process? (1)

b) Which letter describes what happens to the edible oils? (1)

c) What type of compounds are the oils before reaction with hydrogen? (1)

d) What type of compounds are the oils after reaction with hydrogen? (1)

e) What physical property of the oils is changed by this process? (1)

TEST 9
Carbon and its Compounds

1 **Matching pairs question**. For questions (a) to (c) there are five alternative answers, **A** to **E**. Each letter may be used *once, more than once* or *not at all.* Show your choice by writing the letter opposite the number on your answer sheet.

A

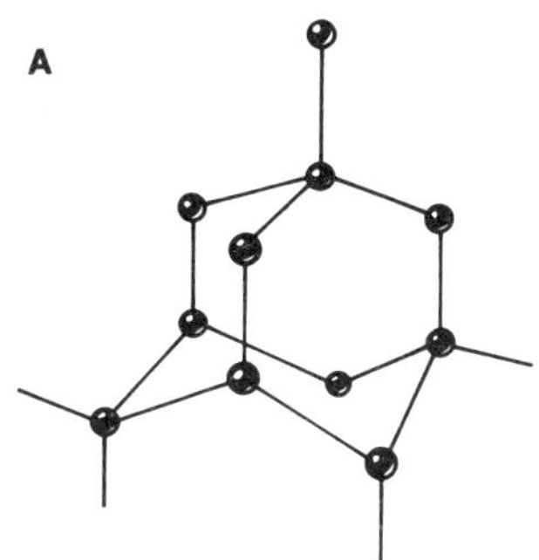

B

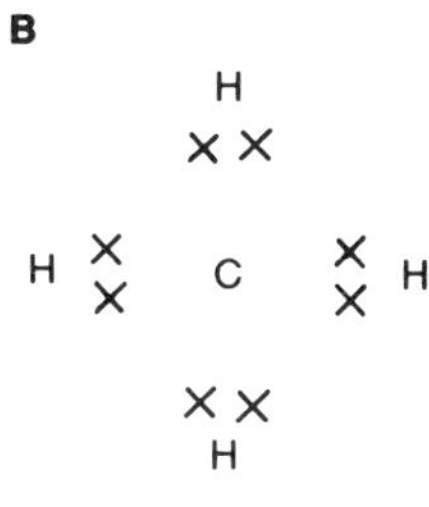

C

O ═ C ═ O

D

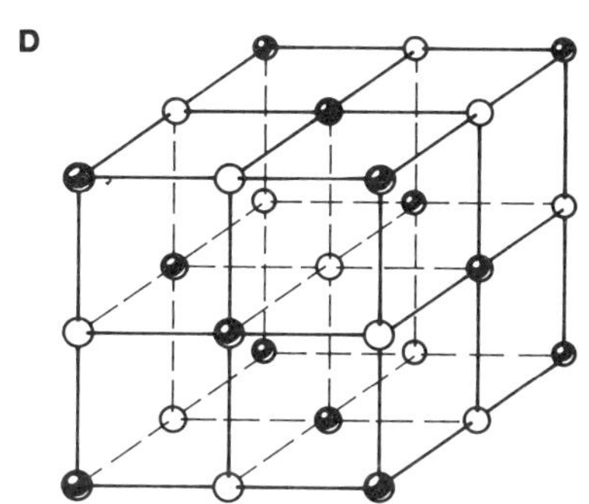

E

Which diagram represents

a) the structure of graphite (1)

b) the structure of diamond (1)

c) the electron arrangement of methane? (1)

2 Copy out this passage and fill in the gaps.

Calcium carbonate (formula = ____________) can occur in three natural forms: limestone, ____________ and ____________. It is used as a raw material in the manufacture of ____________. When it is heated, calcium carbonate decomposes into ____________ and the gas called ____________. (6)

3 What is the electronic structure of carbon atoms?

A **1, 4** **B** **2, 2** **C** **2, 4** **D** **2, 6** **E** **2, 8, 4** (1)

4 a) Give two ways in which carbon dioxide is taken from the atmosphere in the carbon cycle. (2)

b) Give two ways in which carbon dioxide is put into the atmosphere in the carbon cycle. (2)

5 Give the formula for

a) sodium carbonate (1)

b) sodium hydrogencarbonate. (1)

c) How will heating these chemicals show the difference between them? (1)

6 Write a balanced symbol equation showing the reduction of iron(III) oxide with carbon monoxide in a blast furnace. (2)

7 For questions (a) and (b), choose the correct response from the list below.

A **It burns with a blue flame.**

B **It is a good reducing agent.**

C **It contains only the elements carbon and oxygen.**

D **It is a highly poisonous gas.**

E **It is much less dense than air.**

a) Which one of these is *not* a property of carbon monoxide? (1)

b) Which one of these *is* a property of carbon dioxide? (1)

8 Study the following reaction scheme. Give the name or formula for **A, B, C** and **D**.

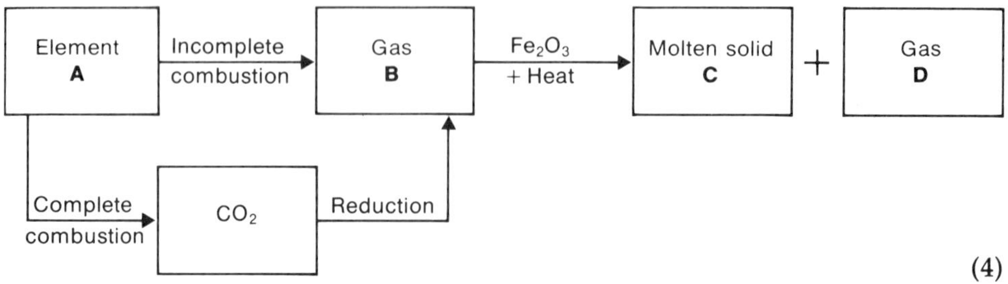

(4)

9 Limewater goes milky when carbon dioxide is bubbled through it.

a) What is limewater a solution of? (1)

b) Write a word equation for the reaction described in the above test. (1)

c) Write a balanced symbol equation for the above reaction. (2)

d) What happens when an excess of carbon dioxide is bubbled through limewater? (1)

TEST 10
Nitrogen and its Compounds

1 a) What percentage of the air is taken up by the gas nitrogen?

A 19 B 21 C 50 D 79 E 89 (1)

b) Why is pure nitrogen often referred to as an *'inert atmosphere'*? (1)

c) How is liquid nitrogen obtained from the air? (2)

2 Here is an illustration of the nitrogen cycle.

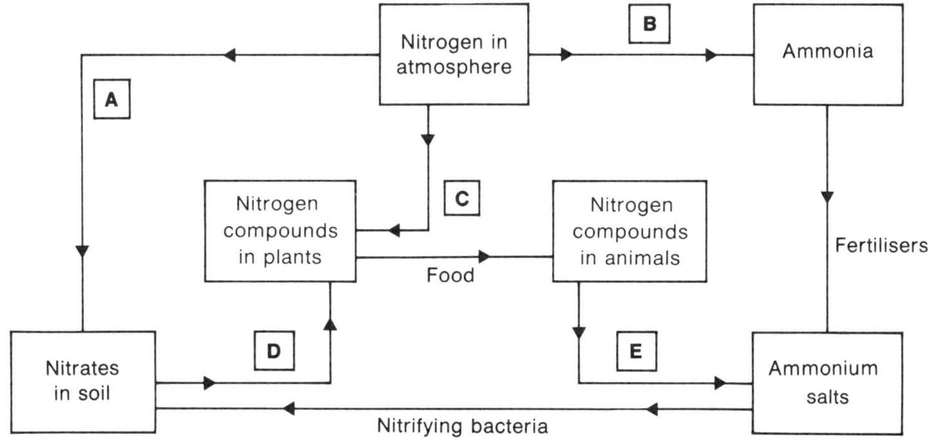

a) What are the processes or conditions labelled **A**, **B**, **C**, **D** and **E**? (5)

b) How can nitrogen return to the atmosphere from plants and animals? (1)

3 **Matching pairs question**. For questions (a) to (c) there are five alternative answers, **A** to **E**. Each letter may be used *once, more than once* or *not at all*. Show your choice by writing the letter opposite the number on your answer sheet.

A **thermal decomposition** **B** **endothermic** **C** **neutralisation**

D **oxidation** **E** **reduction**

Which answer best describes the following reactions?

a) ammonia solution + hydrochloric acid in equal quantities (1)

b) copper metal becoming copper(II) nitrate when added to concentrated nitric acid (1)

c) ammonium chloride dissolving in water (1)

4 Ammonia in water partially splits up into ammonium ions and hydroxide ions. It is a weak alkali and will produce insoluble hydroxides when added to many metal ion solutions.

a) What do you see when a small amount of ammonium hydroxide is added to a solution of Cu^{2+} ions? (1)

b) What do you see when an excess of ammonium hydroxide is added to the solution in (a) above? (2)

5 **The fountain experiment.**

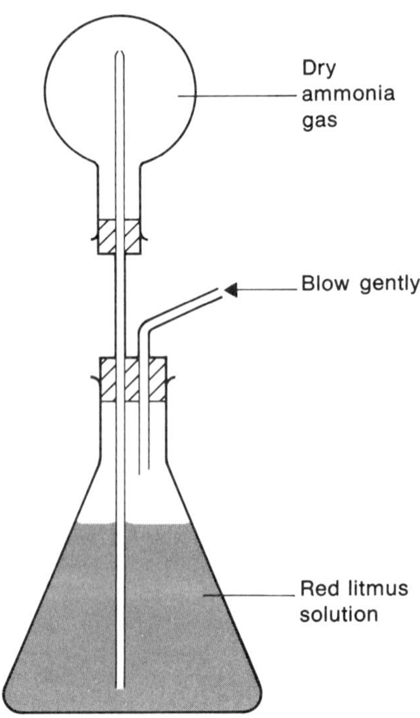

a) Describe what you would *see* happening in this experiment. (3)

b) *Why* does the experiment work? (3)

6 Which one of the following is not a use for ammonia?

A the manufacture of fertilisers

B the manufacture of detergents

C the manufacture of nitric acid

D the manufacture of nylon (1)

7 Some copper(II) nitrate was heated in a test tube. During heating a brown gas was observed and so was a gas which relit a glowing splint. During heating, there was also a colour change in the solid.

a) Name the two gases given off. (2)

b) What was the colour change? (1)

8 Some potassium nitrate was heated in a test tube. Which one of the following observations would be true?

A A brown gas was given off.

B A gas which relit a glowing splint was given off.

C A colour change was seen.

D Potassium oxide was formed.

E Carbon dioxide was produced. (1)

9

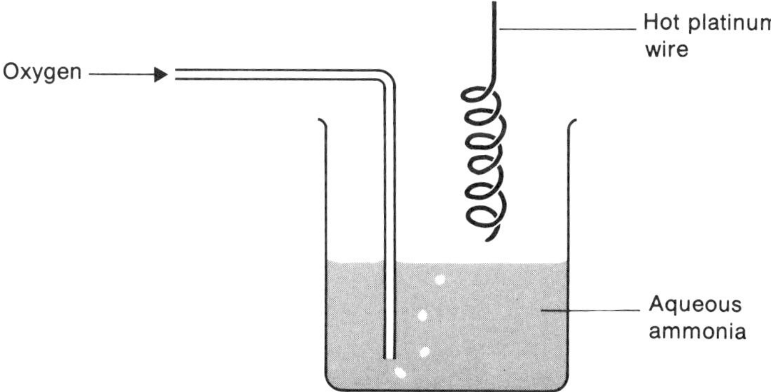

The diagram shows an experiment involving ammonia solution.

a) What kind of reaction does the ammonia undergo? (1)

b) What part does the platinum wire play in the reaction? (1)

c) Why does the wire continue to glow when held in the beaker? (1)

TEST 11
Sulphur and its Compounds

1 Copy out and complete this passage.

Sulphuric acid is a typical acid. With universal indicator added, the colour of the solution will be ____________ . It will react with an alkali to form a salt and ____________ only. The salts of sulphuric acid are called __________ and ____________ . (4)

2 Which *one* of the following is *not* a property of sulphur?

A **It is a non-metal.**

B **It is an impurity in fossil fuels.**

C **It is malleable.**

D **It does not conduct electricity.**

E **It has six electrons in its outer shell.** (1)

3 This sign is seen on a bottle of sulphuric acid. What does it mean?

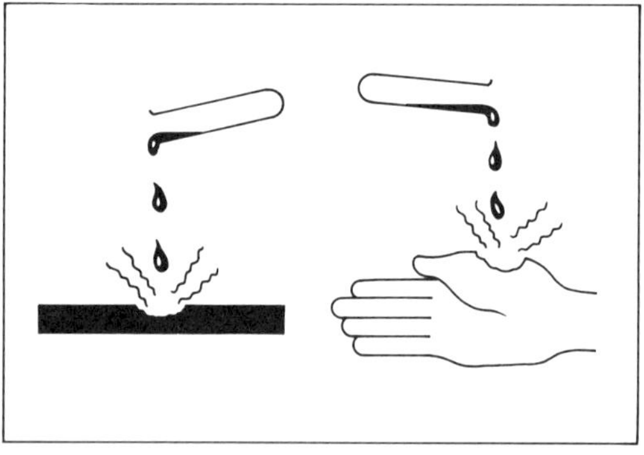

(1)

4 This equation is for burning sulphur in air.

Sulphur + oxygen → sulphur dioxide

a) What colour is the sulphur? (1)

b) What colour is the flame produced as the sulphur burns? (1)

c) Write a balanced symbol equation for the above reaction. (1)

d) What effect does a solution of sulphur dioxide have on universal indicator solution? (1)

5 In Britain there are many considerations as to where to site a sulphuric acid plant. Here are three considerations.

a) easy access to ports

b) easy access to railways or motorways

c) nearby town/city

Say why each of these considerations is important. (3)

6 Which of the following is an important source of sulphur?

A **sea water**

B **atmosphere**

C **impure natural gas**

D **rain water**

E **limestone** (1)

7

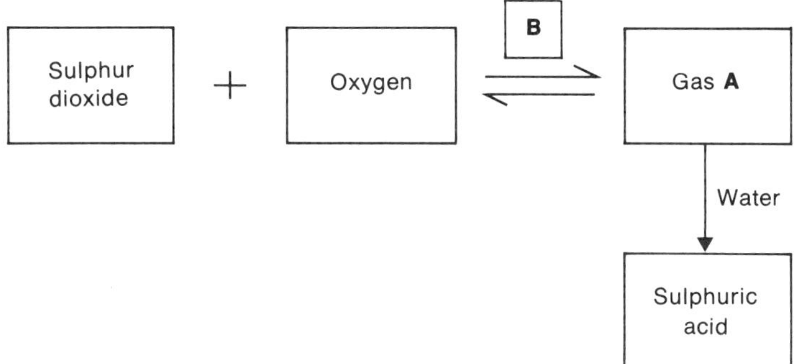

a) Give the name *and* formula of gas **A**. (2)

b) What does the symbol $\rightleftharpoons$ mean? (1)

c) What is reaction **B** known as? (1)

d) What catalyst is used in reaction **B**? (1)

e) Write a balanced symbol equation for the conversion of gas **A** into sulphuric acid by adding water. (2)

f) Why is this reaction not used to produce sulphuric acid in the industrial process? (1)

8 Describe what happens when concentrated sulphuric acid is added to

a) blue copper(II) sulphate crystals (1)

b) sugar. (1)

c) What property of sulphuric acid do these reactions show? (1)

9 Give one commercial use for sulphuric acid. (1)

10 Study the following results table.

Compound	*Colour of aqueous solution*	*Addition of barium chloride solution*	*Other information*
A	Blue	White precipitate	Flame test: blue–green
B	Colourless	White precipitate	pH = 1
C	Colourless	White precipitate	Flame test: yellow
D	Colourless	White precipitate	Flame test: lilac

Give the most likely name or formula for compounds **A**, **B**, **C** and **D**. (4)

TEST 12
Chlorine and the Halogens

1 Copy and complete this table.

Name of halogen	*State at room temperature*	*Colour of halogen*	*Electronic structure*
Fluorine		Yellow–green	2.7
Chlorine	Gas	Yellow–green	
	Liquid	Red–brown	2.8.18.7
Iodine			2.8.18.18.7

(5)

2

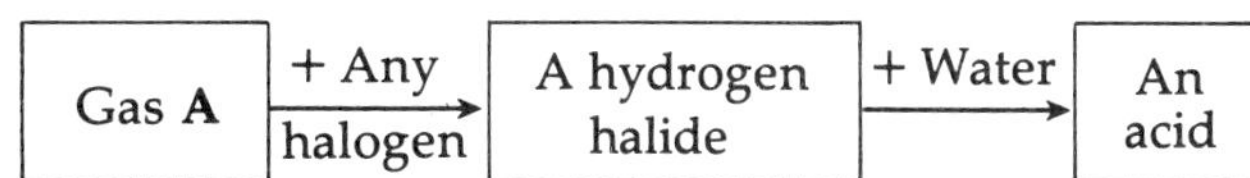

a) What is gas **A**? (1)

b) Which halogen reacts most vigorously with gas **A**? (1)

c) Which of the first four halogens will make the weakest acid? (1)

3 Which halogen will displace chloride ions from solution? (1)

4 In the reaction

$$2FeCl_2(s) + Cl_2(g) \rightarrow 2FeCl_3(s)$$

a) What colour change is seen in the iron chlorides? (1)

b) Is the chlorine an oxidising or reducing agent? (1)

5 a) There are four mistakes in the diagram below. List them. (4)

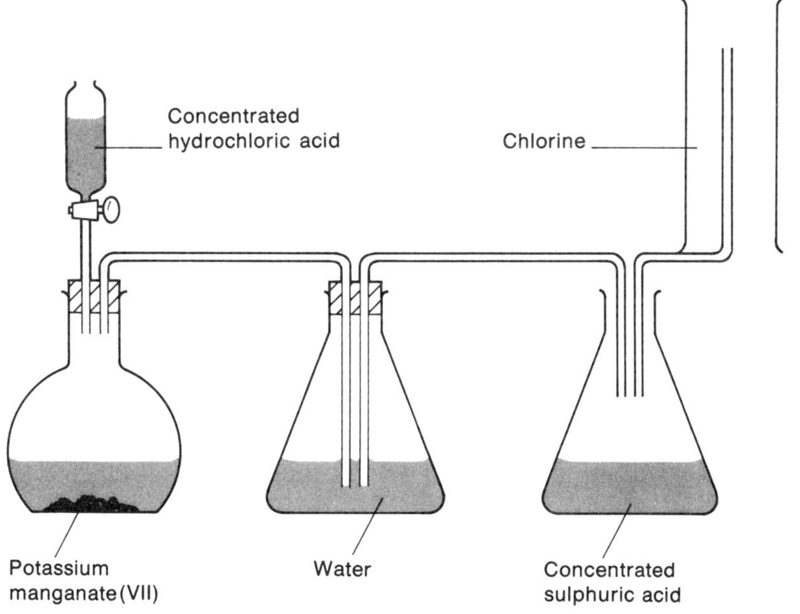

b) Why is the gas passed through water? (1)

c) Why is the gas passed through sulphuric acid? (1)

6 Chlorine gas is made from chloride ions in solution, e.g.

$$2Cl^-(aq) \rightarrow Cl_2(g) + 2e^-$$

a) Name one compound which will produce chloride ions when dissolved in water. (1)

b) Are the chloride ions in the equation above, oxidised or reduced? (1)

7 There are two kinds of chlorine atom commonly found in nature.

$$^{35}_{17}Cl \qquad ^{37}_{17}Cl$$

a) Why are their atomic numbers the same? (1)

b) Why are their mass numbers different? (1)

c) What name is given to atoms of an element which differ in this way? (1)

8 Aluminium chloride is produced when aluminium is heated in chlorine gas. Write a balanced symbol equation for this reaction. (2)

9

Chloroethene (vinyl chloride)	Heat + initiator →	Polychloroethene (polyvinyl chloride)

a) What is polychloroethene (polyvinyl chloride) commonly called? (1)

b) Give *one* use for this substance. (1)

c) What type of reaction is shown above? (1)

d) What does the initiator do? (1)

e) What danger is involved in having objects made of polychloroethene in the home? (2)

TEST 13
A First Test on Metals

1 Which *one* of the following is a metal?

A **aluminium** **B** **hydrogen** **C** **nitrogen** **D** **oxygen**
E **sulphur** (1)

2 Which *one* of the following is *not* a metal?

A **copper** **B** **iron** **C** **lead** **D** **phosphorus**
E **sodium** (1)

3 When copper is heated in air

a) what is its colour before heating? (1)

b) what is its colour after heating? (1)

c) name the new compound which has been made in this reaction. (1)

4 Name the gas given off when zinc reacts with dilute sulphuric acid. (1)

5 The reaction between sodium and dilute sulphuric acid is not one you will see in the school laboratory. Why not? (1)

6 Which *one* of the following is a transition metal?

A **aluminium** **B** **calcium** **C** **copper** **D** **lead**
E **potassium** (1)

7 Copy out and complete the following sentence.

Metals may be identified by the fact that they form ____________ ions by ____________ of one or more electrons when they form ionic compounds with non-metals. (2)

8 a) What do you *see* happening when potassium reacts with water? (2)

b) Give a word equation for this reaction. (1)

c) Give a balanced symbol equation for this same reaction. (2)

9 Iron will displace copper from copper(II) sulphate solution.

a) What colour *change* would you *see* in the solution? (1)

b) Give a word equation for this reaction. (1)

c) Give a balanced symbol equation. (1)

d) Give a balanced ionic equation. (2)

10 Explain in terms of their electronic structures why potassium is more reactive than lithium in water. (2)

11 Which *one* of the following is *correct* for the activity series of metals from most reactive to least reactive?

A **calcium, magnesium, iron, sodium, potassium**

B **magnesium, potassium, sodium, calcium, iron**

C **potassium, sodium, calcium, magnesium, iron**

D **sodium, potassium, magnesium, iron, calcium**

E **iron, calcium, magnesium, potassium, sodium** (1)

12 Explain in terms of the arrangement of atoms in the structure of a piece of metal

a) how metals conduct electricity (2)

b) why metals are malleable. (2)

13 You are given some aluminium sulphate solution.

a) Which *one* of the following metals would you use if you wanted to displace aluminium from the solution?

A **copper**

B **iron**

C **lead**

D **magnesium**

E **silver** (1)

b) Give reasons for your answer to part (a). (2)

TEST 14

A Second Test on Metals

---

1 Iron is extracted from its ores in the blast furnace. Write on to your answer paper the letters **A**, **B**, **C**, **D** and **E**, and name the raw materials or products they represent in the diagram on the opposite page.

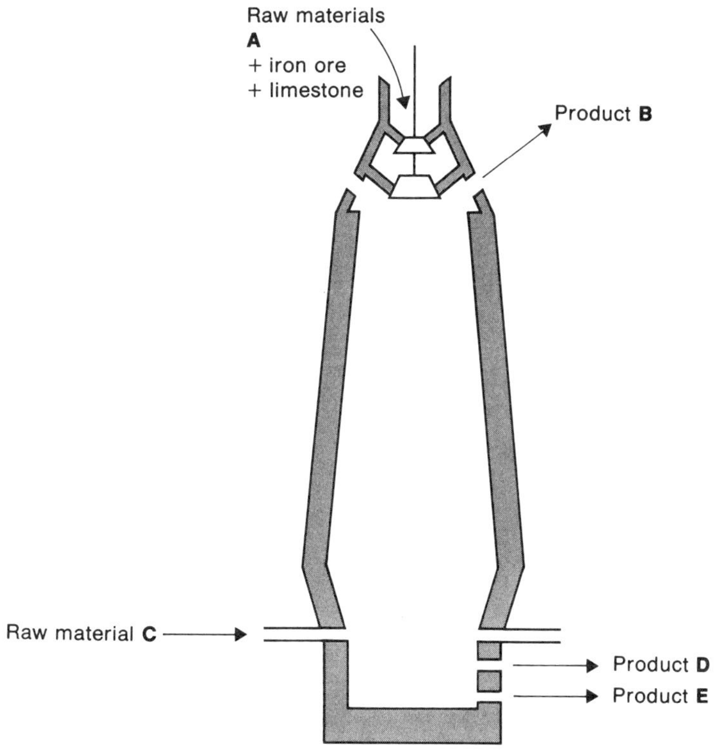

(5)

2 a) Which element, present in the iron as an impurity, is first removed and then added again in the production of most types of steel? (1)

b) What disadvantage does iron have which explains why we usually use steel rather than iron? (1)

3 Which *one* of the following is present in stainless steel?

A **calcium** B **copper** C **chromium** D **silver**
E **sulphur** (1)

4 Duralumin is used in kitchen utensils and aircraft in preference to pure aluminium. Why is this?

A **It is lighter.**

B **It is stronger.**

C **It is coloured.**

D **It is made more easily.**

E **It is cheaper.** (1)

5 This question concerns the effect of heat on copper(II) carbonate.

a) What colour is copper(II) carbonate before heating? (1)

b) What is the colour of the chemical left in the test tube after heating? (1)

c) Give a word equation for this reaction. (1)

d) Give a balanced symbol equation for this reaction. (2)

6 Copy out the following table and fill in the blanks.

Metal	*Ore*	*Method of extraction*
Aluminium		
	Haematite	

(2) ($\frac{1}{2}$ mark each)

7 **Matching pairs question**. For questions (a) to (d) each question has five alternative answers, **A** to **E**. Each letter may be used *once, more than once* or *not at all.* Show your choice by writing the correct letter (**A** to **E**) at the side of the question number on your answer sheet. From the list

A **car bodies**

B **joining wires together**

C **aircraft wings**

D **cutlery**

E **ornaments**

choose the most popular use for the alloys

a) stainless steel (1)

b) mild steel (1)

c) brass (1)

d) solder. (1)

8 The recently discovered wonder element batlium will displace zinc from a solution of zinc nitrate. Magnesium will displace batlium from a solution of batlium nitrate. Write the three metals in order from most reactive to least reactive. (1)

9 Study the flow diagram on the opposite page, and then answer the questions.

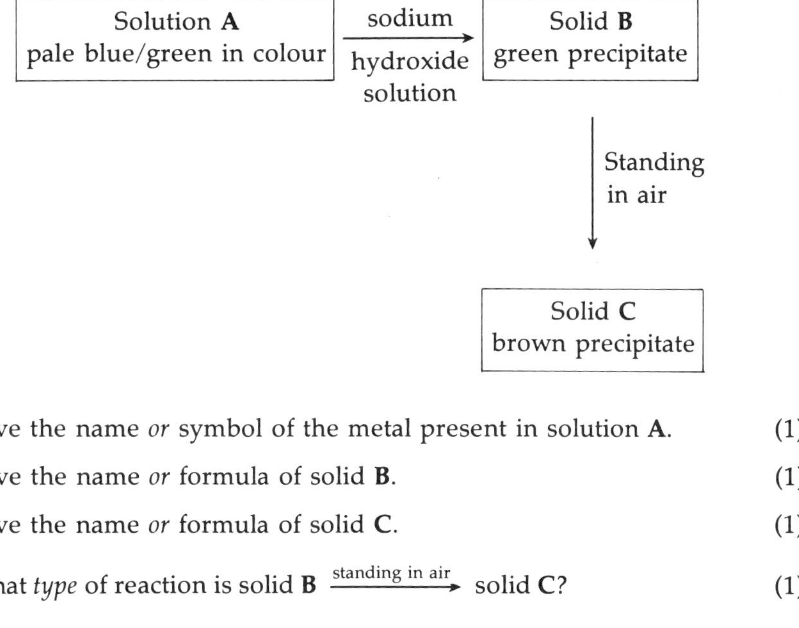

a) Give the name *or* symbol of the metal present in solution **A**. (1)

b) Give the name *or* formula of solid **B**. (1)

c) Give the name *or* formula of solid **C**. (1)

d) What *type* of reaction is solid **B** $\xrightarrow{\text{standing in air}}$ solid **C**? (1)

10 Aluminium hydroxide is said to be amphoteric. What does this mean? (1)

11 Explain why aluminium and iron require different methods of extraction. (2)

12 Aluminium and iron both oxidise in air. Why is this a problem with iron, but an advantage with aluminium? (2)

TEST 15
A First Test on Organic Chemistry

1 Copy and complete this paragraph.

Organic chemistry is the study of chemicals which contain atoms of the element ____________ . Compounds which are not considered to be organic include the oxides of this element, namely ____________ ____________ and ____________ ____________ . (3)

2 Write out or draw the correct answers for **A** to **F**.

Name of alkane	*Molecular formula*	*Structural formula*
Methane	CH_4	**A**
B	**C**	H–C(H)(H)–C(H)(H)–H
Propane	**D**	**E**
Butane	**F**	H–C(H)(H)–C(H)(H)–C(H)(H)–C(H)(H)–H

(6)

3 What is the major gas present in North Sea gas?

A methane **B** ethane **C** ethene
D ethyne **E** propane (1)

4 Which of the following processes is the breaking down of long-chain hydrocarbons into short-chain hydrocarbons?

A cracking **B** electrolysis **C** ionisation
D oxidation **E** polymerisation (1)

5 What name is given to a straight-chained alkene with three carbon atoms?

A methane **B** ethane **C** ethene
D propane **E** propene (1)

6 a) What is meant by the term 'unsaturated'? (1)

b) Name one unsaturated organic compound. (1)

7 What happens to bromine water when it is shaken with a sample of an alkene? (1)

8 Name the two products from the efficient burning of methane. (2)

9 a) What type of bonding occurs in the methane molecule? (1)

b) Sketch the electron arrangement (outer energy levels only) of the methane molecule. (1)

10 Crude oil can be distilled into fractions. Below is a table of data on four liquid fractions.

Name of fraction	*Range of carbon atoms in each molecule*	*Range of boiling points (°C)*
Petrol	4–12	40–75
Kerosine	9–16	75–150
Diesel oil	15–25	220–250
Lubricating oil	20–70	250–350

a) Why is the fraction containing 1–3 carbon atoms not included in this table? (1)

b) Give one use for each of these four fractions. (4)

c) Which of the above fractions will boil to dryness when placed in the test tube, as shown in the diagram below? (1)

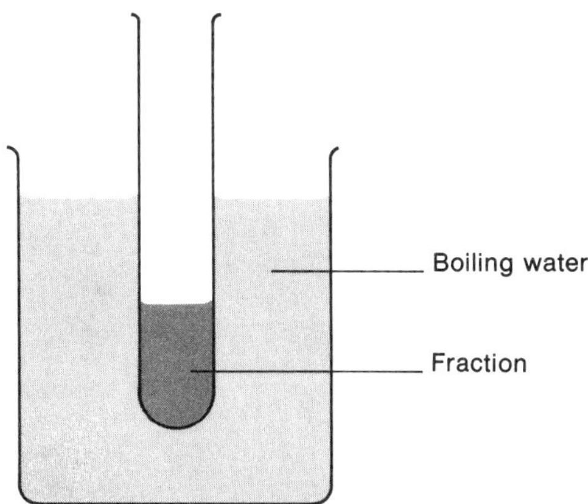

11 Chlorine and methane were sealed in a vessel and irradiated with light. When the results were analysed there were four chlorine-substituted carbon compounds and a gas which turned universal indicator paper red. Give the name or formula of each of the five products. (5)

TEST 16
A Second Test on Organic Chemistry

1 What are the best conditions for producing ethanol by fermentation with yeast?

A **water at 30 °C**

B **sugar at 100 °C**

C **water and sugar at 100 °C**

D **water at 100 °C**

E **water and sugar at 30 °C** (1)

2 For questions (a) and (b), choose the correct response from the list below.

A **alcohols** B **alkalis** C **alkanes** D **carboxylic acids**

Which of the above groups is likely to produce

a) an aqueous solution of pH4 (1)

b) an ester when reacted with ethanoic acid? (1)

3 Name the chemical that can be added to propene to make propanol. (1)

4 Give a property of thermoplastics and a property of thermosetting plastics to show the difference between them. (2)

5 The diagram below shows the apparatus needed to prepare an alkene.

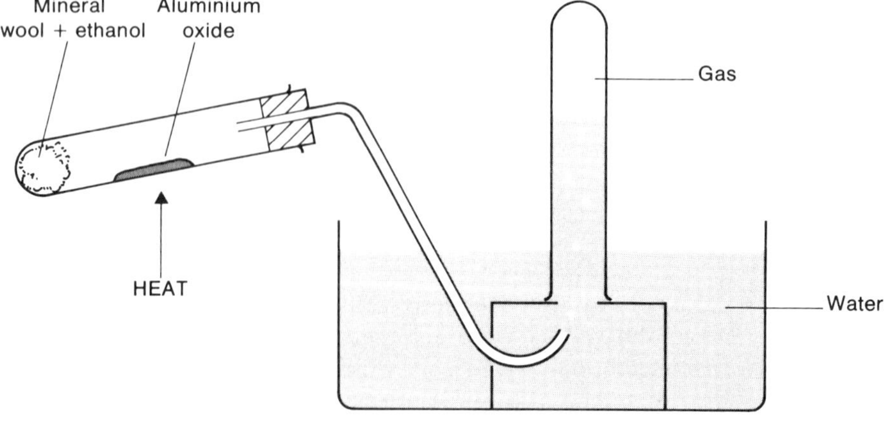

a) Name the alkene produced. (1)

b) Why is the first tube of gas which is collected thrown away? (1)

c) How would you show that the gas produced is an alkene? (1)

d) What name is given to the process shown in the diagram?

A **dehydration** **B** **dehydrogenation** **C** **deoxidation**

D **redox** **E** **reduction** (1)

6

Glucose $C_6H_{12}O_6$	$\xrightarrow[\text{(in yeast)}]{\text{Zymase}}$	Ethanol C_2H_5OH	+	Carbon dioxide CO_2

a) Write a balanced symbol equation for this reaction. (2)

b) How many moles of ethanol can be made from one mole of glucose? (1)

c) Name one alcohol other than ethanol. (1)

d) What kind of chemical is zymase? (1)

7 The diagram below is of apparatus needed to carry out an experiment with ethanol and potassium dichromate(VI).

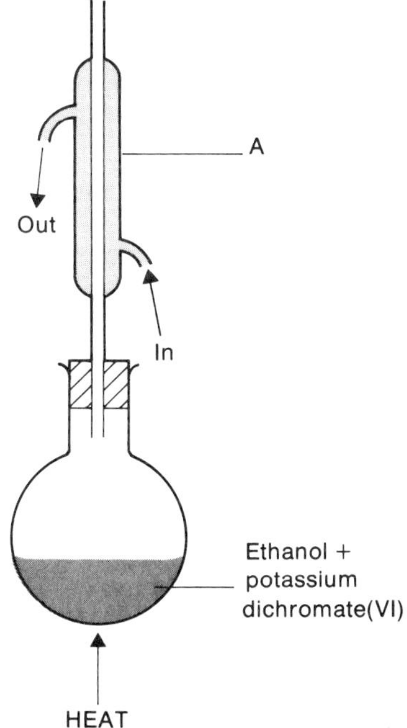

Look at the diagram on page 41.

a) Name the apparatus **A**. (1)

b) Why does the ethanol not evaporate from the flask? (1)

c) Name the main product of the reaction. (1)

d) What is the name of this process?

A **cracking**

B **distillation**

C **refluxing**

D **saponification**

E **subliming** (1)

8 This diagram represents a molecule of sodium stearate.

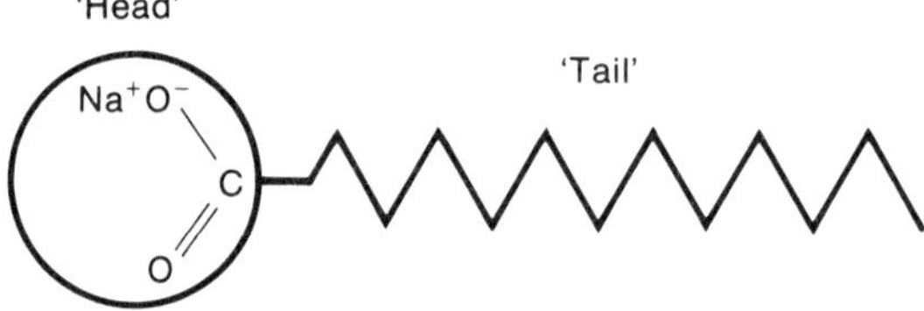

a) What is sodium stearate used for? (1)

b) Why is the head end described as hydrophilic? (1)

c) What similar word describes the tail end? (1)

d) The tail consists of atoms of *two* elements. Name them. (1) ($\frac{1}{2}$ each)

e) How does this molecule help in removing grease and dirt from materials? (3)

9 a) What is the name of the ester formed by the reaction of ethanoic acid and ethanol? (1)

b) Concentrated sulphuric acid is added to remove the second product of this reaction. What is the second product? (1)

c) Give two uses of esters. (2)

TEST 17
Electrolysis

1 The diagram below shows the electrolysis of dilute sulphuric acid.

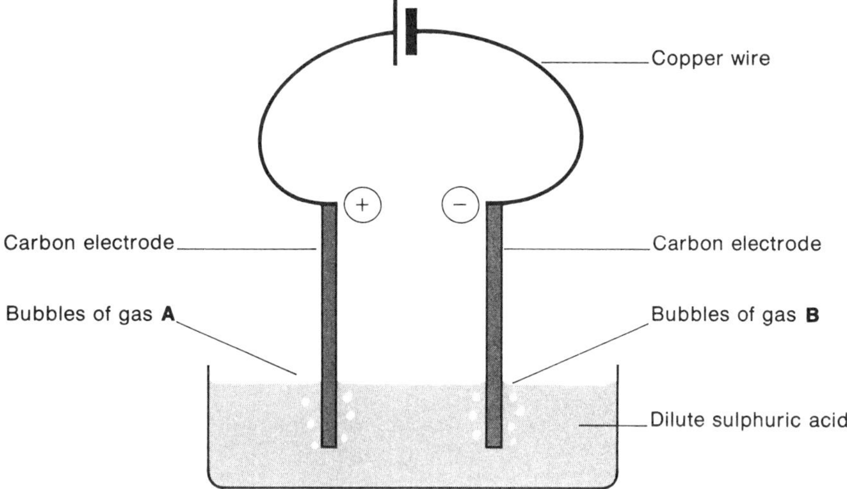

a) What is the correct name for the positive electrode? (1)

b) What is gas **A**? (1)

c) What is gas **B**? (1)

d) What particles carry the electric current through the electrolyte? (1)

e) What particles carry the electric current through the copper wire? (1)

2 Which *one* of the following is *never* discharged at the positive electrode during electrolysis?

A **bromine**

B **chlorine**

C **hydrogen**

D **iodine**

E **oxygen** (1)

3 **Matching pairs question**. For questions (a) to (d) each question has five alternative answers, **A** to **E**. Each letter may be used *once, more than once* or *not at all*. Show your choice by writing the correct letter (**A** to **E**) at the side of the question number on your answer sheet. From the list

A **aluminium**

B **copper**

C **hydrogen chloride**

D **iodine**

E **oxygen**

choose the chemical which

a) forms a divalent negative ion (1)

b) may be anodised (1)

c) forms an electrolyte when dissolved in pure water (1)

d) may be used in electrical wires and forms a divalent ion. (1)

4 List, using symbols and charges, *all* the ions present in a solution of sodium chloride in water. (2)

5 This question involves three chemicals, **A**, **B** and **C**, from the following list.

aluminium oxide **carbon** **copper(II) oxide** **lead(II) oxide**

sodium carbonate **sodium chloride** **sulphur**

The table below gives information on the three chemicals, **A**, **B** and **C**.

Substance	*Appearance of solid*	*Conduction of electricity*		*Products at electrodes when molten*	
		Molten	*In water*	+	−
A	Dark grey powder	Yes	Insoluble	None	None
B	Yellow powder	No	Insoluble	None	None
C	White crystals	Yes	Yes	Chlorine	Sodium

a) Use the list and table above to suggest possible identities for **A**, **B** and **C**. (3)

b) What type of bonding is present in **C**? (1)

6 Use the five simple cells, **A** to **E**, shown at the top of the opposite page to answer the following questions.

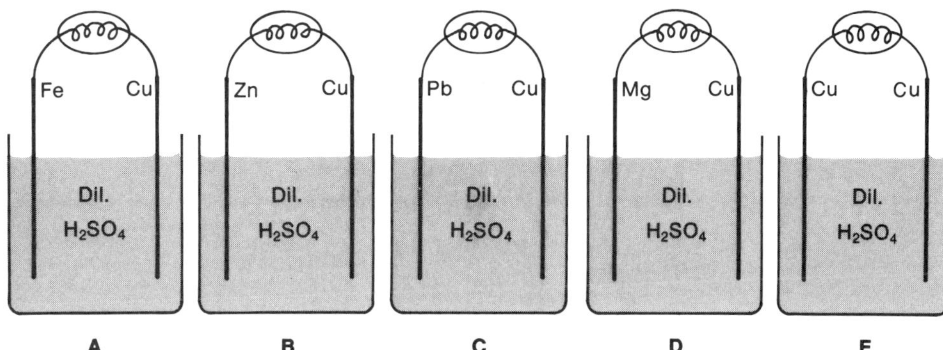

a) Which cell will produce the brightest light? (1)

b) Which cell will produce no light? (1)

c) Which cell will produce the least light for the longest time? (1)

d) Explain your answer to (a) above. (2)

e) Explain your answer to (b) above. (2)

7 Explain why an electrovalent (ionic) compound conducts electricity when molten, but not when it is solid. (2)

8 The electrolysis of molten sodium chloride is illustrated in the following diagram.

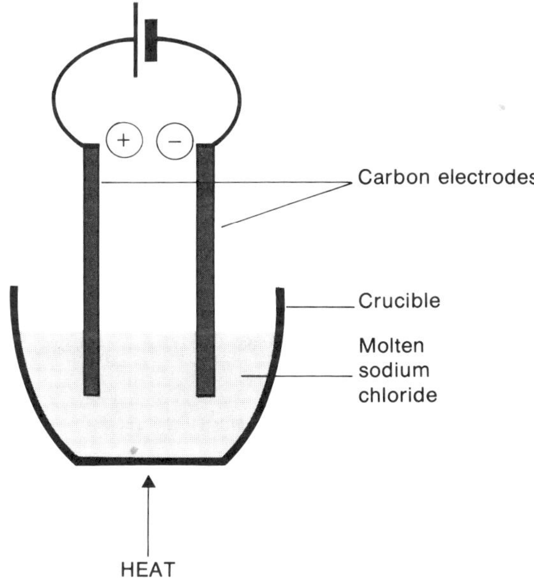

Write balanced symbol equations for the discharge of the ion at

a) the positive electrode (2)

b) the negative electrode. (2)

c) How may the melting point of sodium chloride be lowered? (1)

TEST 18
Oxidation and Reduction

1 For the following reactions, say whether the substance underlined has been oxidised or reduced.

a) <u>Calcium</u> + oxygen → calcium oxide (1)

b) <u>Carbon</u> monoxide + oxygen → carbon dioxide (1)

c) <u>Iron</u>(III) oxide + carbon monoxide → iron + carbon dioxide (1)

2 Copper(II) oxide can be reduced by heating in a stream of dry hydrogen gas.

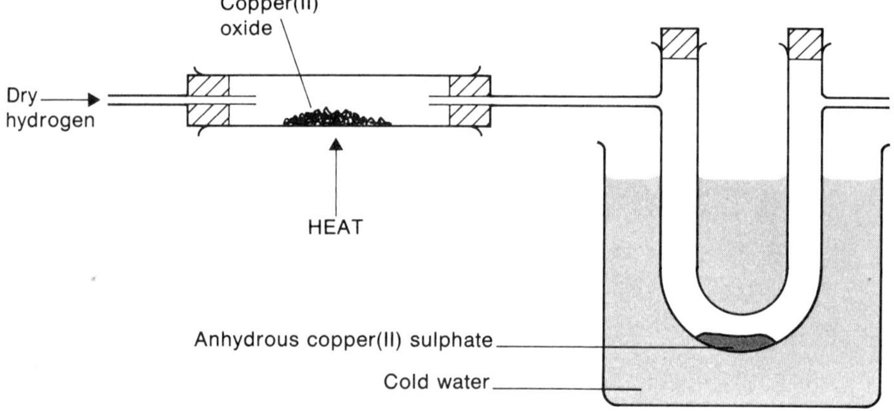

a) What colour *change* occurs in the copper(II) oxide? (1)

b) Write a word equation for the reaction of copper(II) oxide and hydrogen. (1)

c) Write a symbol equation for the reaction in (b). (1)

d) What colour change occurs in the U tube? (1)

e) If the copper(II) oxide has been reduced, what has been oxidised? (1)

3 Copy out and complete this paragraph.

Oxidation can be thought of in terms of gaining atoms of __________ (e.g. when magnesium burns in air, it forms __________ __________). The reverse is also true. Losing these atoms means that something has been __________ (e.g. when lead oxide is heated with carbon to form __________). (4)

4 For the following reactions, say whether the substance underlined is oxidised or reduced.

a) Magnesium + <u>nitrogen</u> → magnesium nitride (1)

b) <u>Iron</u> + chlorine → iron(III) chloride (1)

c) Give a reason for your answer to part (b). (1)

5 Lead has two valency states – lead(II) and lead(IV). The ions are Pb^{2+} and Pb^{4+}, respectively.

A **Pb^{2+} may be changed to Pb^{4+}**

B **Pb^{4+} may be changed to Pb^{2+}**

a) Which reaction (**A** or **B**) involves a loss of electrons? (1)

b) Which reaction shows oxidation? (1)

c) Which reaction needs a reducing agent in order for it to work? (1)

6 In displacement reactions, metals high in the activity series displace metal ions low in the series. For the following ionic reactions, write out the *name* of the metal which is reduced.

a) $Zn^{2+}(aq) + Mg(s) \rightarrow Zn(s) + Mg^{2+}(aq)$ (1)

b) $2Al^{3+}(aq) + 3Ca(s) \rightarrow 2Al(s) + 3Ca^{2+}(aq)$ (1)

7 For the following reaction

Potassium iodide + bromine → potassium bromide + iodine

a) Which substance is reduced? (1)

b) Which substance is the oxidising agent? (1)

8 For questions (a) and (b), choose the correct response from the list below.

A **aluminium oxide**

B **calcium oxide**

C **magnesium oxide**

D **potassium oxide**

E **silver oxide**

Which of the above oxides

a) can be reduced by hydrogen (1)

b) cannot be reduced by the metal sodium? (1)

9 Potassium dichromate(VI) is an oxidising agent. Name the product formed from the following, when treated with potassium dichromate(VI).

a) ethanol (1)

b) $Fe^{2+}(aq)$ (1)

10

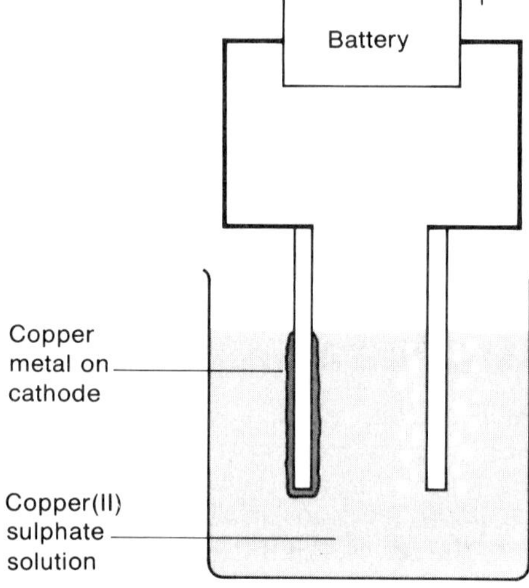

a) Name the gas at the anode. (1)

b) Write an ionic equation showing how the copper ions at the cathode become copper metal. (2)

c) Have the copper ions been oxidised or reduced? (1)

TEST 19
A First Test on Symbols, Formulae and Equations

1 Use the following table of symbols and valencies to help answer questions (a) and (b).

Valency 1		*Valency 2*		*Valency 3*
Sodium Na Potassium K Silver Ag	Hydrogen-carbonate HCO_3 Chloride Cl Bromide Br Hydroxide OH	Calcium Ca Magnesium Mg Iron(II) Fe Copper(II) Cu Barium Ba	Oxide O Carbonate CO_3 Sulphate SO_4	Aluminium Al Iron(III) Fe

a) Give the correct formula for each of the following chemicals.

i)	sodium bromide	ii)	potassium hydroxide
iii)	silver chloride	iv)	barium chloride
v)	magnesium hydroxide	vi)	calcium hydrogencarbonate
vii)	iron(II) oxide	viii)	sodium carbonate

(4) ($\frac{1}{2}$ mark each)

b) What are the names of the following compounds?

i)	$CaCO_3$	ii)	$MgCl_2$	iii)	Fe_2O_3	iv)	$CuSO_4$
v)	$AlBr_3$	vi)	Ag_2O	vii)	$Fe(OH)_2$	viii)	$Ba(HCO_3)_2$

(4) ($\frac{1}{2}$ mark each)

2 **Matching pairs question**. For questions (a) to (e), there are six alternative answers, **A** to **F**. Each letter may be used *once, more than once* or *not at all*. Show your choice by writing the letter opposite the number on your answer sheet. From the list

A	$\mathbf{(NH_4)_3PO_4}$	B	$\mathbf{FeI_3}$	C	$\mathbf{CaI_2}$
D	$\mathbf{NH_4NO_3}$	E	$\mathbf{Cu(NO_3)_2}$	F	$\mathbf{FeI_2}$

choose the letter which represents

a) ammonium nitrate (1)

b) ammonium phosphate (1)

c) iron(III) iodide (1)

d) copper(II) nitrate (1)

e) calcium iodide. (1)

3 Give the correct formula for 1 molecule of each of the following elements.

a)	oxygen	b)	hydrogen	c)	nitrogen	d)	chlorine
e)	helium	f)	bromine				

(3) ($\frac{1}{2}$ mark each)

4 Give the correct formula, for each of the following compounds.

a)	ammonia	b)	hydrogen chloride
c)	concentrated sulphuric acid	d)	sulphur dioxide
e)	carbon dioxide	f)	zinc hydroxide
g)	potassium manganate(VII)	h)	hydrogen peroxide

(4) ($\frac{1}{2}$ mark each)

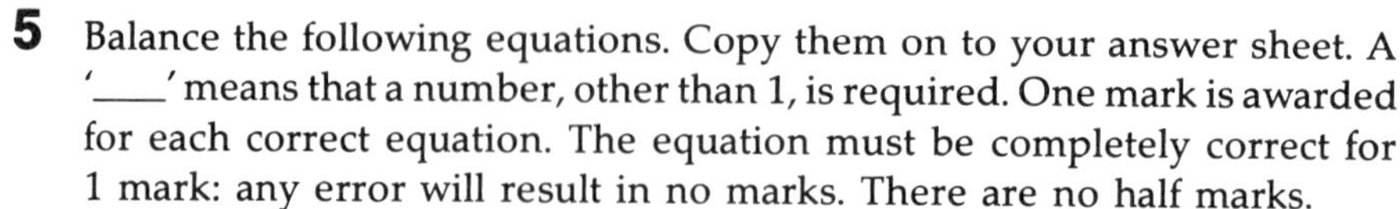

5 Balance the following equations. Copy them on to your answer sheet. A '____' means that a number, other than 1, is required. One mark is awarded for each correct equation. The equation must be completely correct for 1 mark: any error will result in no marks. There are no half marks.

a) ____ $Cu + O_2 \rightarrow$ ____ CuO (1)

b) ____ $H_2 + O_2 \rightarrow$ ____ H_2O (1)

c) N_2 + ____ $H_2 \rightleftharpoons$ ____ NH_3 (1)

d) ____ $Na + O_2 \rightarrow$ ____ Na_2O (1)

e) ____ $NaOH + H_2SO_4 \rightarrow Na_2SO_4$ + ____ H_2O (1)

f) ____ Al + ____ $Cl_2 \rightarrow$ ____ $AlCl_3$ (1)

g) ____ P + ____ $O_2 \rightarrow$ ____ P_2O_5 (1)

h) ____ Fe + ____ $H_2O \rightarrow Fe_3O_4$ + ____ H_2 (1)

6 Which one of the following is correct for the formula of calcium phosphate?

A $CaPO_4$ **B** Ca_2PO_4 **C** Ca_3PO_4 **D** $Ca(PO_4)_2$

E $Ca_3(PO_4)_2$ (1)

7 Which one of the following is correct for the formula of aluminium oxide?

A Al_3O_2 **B** Al_2O_3 **C** Al_3O_3 **D** Al_2O_4 **E** Al_3O_4 (1)

TEST 20

A Second Test on Symbols, Formulae and Equations

---

1 Name each of the following.

a) Ca b) C c) CO d) CO_2

e) CaO f) CO_3^{2-} g) $Ca(OH)_2$ h) Ca^{2+}

(4) ($\frac{1}{2}$ mark each)

2 Explain the meaning of the symbols (s), (l), (aq) and (g) in the following equation, with reference to *each chemical*.

$$Na_2CO_3(s) + 2HCl(aq) \rightarrow 2NaCl(aq) + H_2O(l) + CO_2(g)$$

($2\frac{1}{2}$) ($\frac{1}{2}$ mark each)

3 **Matching pairs question**. For questions (a) to (g) there are eight alternative answers, **A** to **H**. Each letter may be used *once, more than once* or *not at all*. Show your choice by writing the letter opposite the number on your answer sheet. From the list

A	**$HCl(aq)$**	B	**$H_2O(s)$**
C	**$FeSO_4(s)$**	D	**$Fe_2(SO_3)_3(s)$**
E	**$H_2O(l)$**	F	**$HCl(g)$**
G	**$FeSO_3(s)$**	H	**$FeS(s)$**

choose the formula which represents a molecule of

a) hydrogen chloride gas (1)

b) water at +20 °C (1)

c) iron(II) sulphide solid (1)

d) iron(III) sulphite solid (1)

e) hydrochloric acid (1)

f) iron(II) sulphate solid (1)

g) water at −20 °C. (1)

4 Write balanced symbol equations for each of the following.

a) Hydrated copper(II) sulphate $\underset{\text{cool}}{\overset{\text{heat}}{\rightleftharpoons}}$ anhydrous copper(II) sulphate + water (1)

b) Ammonium chloride $\rightleftharpoons$ ammonia + hydrogen chloride (1)

5 Give symbols for each of the following.

a) an ammonium ion ($\frac{1}{2}$)

b) a magnesium ion ($\frac{1}{2}$)

c) a hydroxide ion (hydroxyl ion) ($\frac{1}{2}$)

6 Balance the following equations. A '____' means that a number, other than 1, is needed. There is one mark for a correct answer: there are no half marks.

a) $CH_4 +$ ____ $O_2 \rightarrow CO_2 +$ ____ H_2O (1)

b) $C_2H_5OH +$ ____ $O_2 \rightarrow$ ____ $CO_2 +$ ____ H_2O (1)

7 **Matching pairs question**. For questions (a) to (c) there are six alternative answers, **A** to **F**. Each letter may be used *once, more than once* or *not at all.* Show your choice by writing the letter opposite the number on your answer sheet. From the list

A	C_2H_4	B	C_2H_6	C	C_3H_8
D	CH_3COOH	E	C_2H_5OH	F	C_4H_{10}

choose the formula which represents a molecule of

a) butane (1)

b) ethene (1)

c) ethanol. (1)

8 Write an *ionic equation* for each of the following reactions.

a) Silver nitrate solution + sodium chloride solution → silver chloride + silver nitrate solution (2)

b) Sodium sulphate solution + barium nitrate solution → sodium nitrate solution + barium sulphate (2)

c) Dilute hydrochloric acid + potassium hydroxide solution → potassium chloride solution + water (2)

d) Hydrogen ions formed at the cathode, in the electrolysis of dilute sulphuric acid, produce hydrogen gas. (2)

TEST 21
A First Test on Calculations and Graphs

You will need the following information for this test.

Approximate relative atomic masses

Element	*Symbol*	*Relative atomic mass*
Hydrogen	H	1
Carbon	C	12
Nitrogen	N	14
Oxygen	O	16
Sodium	Na	23
Sulphur	S	32
Calcium	Ca	40
Iron	Fe	56
Copper	Cu	63.5

You may use calculators. You will need a sheet of graph paper for question 6.

1 If one mole of sodium atoms has a mass of 23 g, what is the mass of

a) 4 moles of sodium atoms (1)

b) 0.1 mole of sodium atoms? (1)

2 If 56 g is the mass of one mole of iron atoms, how many moles are in

a) 280 g of iron atoms (1)

b) 0.56 g of iron atoms? (1)

3 Calculate the molar masses of the following compounds.

a) sodium oxide (Na_2O) (1)

b) iron(III) nitrate ($Fe(NO_3)_3$) (1)

c) carbon dioxide (1)

d) sulphuric acid (1)

4 Below are the results from an experiment carried out to find the formula of an oxide of copper. In this experiment some copper metal was weighed before and after heating to constant mass. This makes sure that all the copper is converted to the oxide. The heating takes place in a crucible.

Mass of the crucible = 13.80 g

Mass of the crucible + copper metal = 20.15 g

Mass of the crucible + copper oxide = 21.75 g

a) How many grams of copper metal were used? (1)

b) How many grams of copper oxide were formed? (1)

c) How many grams of oxygen gas had been used in order to form the copper oxide? (1)

d) Use your answer to part (a) to calculate how many moles of copper metal were used at the start. (1)

e) Use your answer to part (c) to calculate how many moles of oxygen gas had combined with this copper. (1)

f) What is the ratio of copper to oxygen? (1)

g) What is the empirical formula of this oxide of copper? (1)

5 The formula of calcium carbonate is $CaCO_3$.

a) What is the molar mass of calcium carbonate? (1)

b) What percentage, by mass, is taken up by oxygen in calcium carbonate? (1)

6 Use your graph paper to answer this question. Some magnesium metal was placed in an excess amount of dilute hydrochloric acid. The volume of hydrogen gas given off was measured at ten second intervals from the start of the experiment.

Time (s)	0	10	20	30	40	50	60
Volume of H_2 (cm^3)	0	120	175	210	230	240	240

The equation for this experiment is:

$$Mg(s) + HCl(aq) \rightarrow MgCl_2(aq) + H_2(g)$$

a) Draw a graph of volume of hydrogen collected against time. (4)

b) If one mole of any gas at room temperature and pressure takes up 24 litres, how many moles of hydrogen gas were collected? (1)

c) What was the mass, in grams, of the magnesium metal added at the start of the experiment? (1)

7 The formula of sodium carbonate crystals was known to be $Na_2CO_3 \cdot xH_2O$. The following experiment and calculation were designed to find the value of x in this formula. Sodium carbonate crystals were heated to constant mass in order to drive off the water of crystallisation.

Mass of hydrated sodium carbonate crystals = 7.15 g

Mass of anhydrous sodium carbonate = 2.65 g

a) What is the mass, in grams, of water lost? (1)

b) What percentage (to the nearest whole number) is this of the mass of hydrated sodium carbonate crystals?

A **10** B **37** C **41** D **59** E **63** (1)

c) What is the molar mass of anhydrous sodium carbonate (Na_2CO_3)? (1)

d) How many moles of anhydrous sodium carbonate were formed at the end of the experiment? (1)

e) What is the molar mass of water? (1)

f) How many moles of water were lost in the experiment? (1)

g) Using your answers to (d) and (f), work out the value of x in the formula $Na_2CO_3 \cdot xH_2O$. (1)

TEST 22
A Second Test on Calculations and Graphs

You will need the following information for this test.

Approximate relative atomic masses

Element	*Symbol*	*Relative atomic mass*
Hydrogen	H	1
Nitrogen	N	14
Oxygen	O	16
Sodium	Na	23
Magnesium	Mg	24
Sulphur	S	32
Chlorine	Cl	35.5
Calcium	Ca	40
Iron	Fe	56
Zinc	Zn	65
Barium	Ba	137

You may use calculators.

1 When hydrogen and sulphur combine, a gas called hydrogen sulphide is produced.

$$H_2(g) + S(s) \rightarrow H_2S(g)$$

a) How many moles of H_2S are made from one mole of sulphur? (1)

b) What mass of H_2S is made from 32 g sulphur? (1)

2 Sodium carbonate reacts with excess hydrochloric acid.

$$Na_2CO_3(s) + 2HCl(aq) \rightarrow 2NaCl(aq) + CO_2(g) + H_2O(l)$$

a) How many moles of sodium chloride are produced from one mole of sodium carbonate? (1)

b) What mass of sodium chloride is produced from one mole of sodium carbonate? (1)

3 Magnesium nitride is made from magnesium reacting with excess nitrogen.

$$3Mg(s) + N_2(g) \rightarrow Mg_3N_2(s)$$

What mass of magnesium nitride can be made from 36 g magnesium? (2)

4 When sodium reacts with excess water, the gas hydrogen is produced (remember to write an equation). How many grams of hydrogen would be made from the reaction of 11.5 g sodium metal with water? (2)

5 The graph shows the temperature of water as it was heated from −20 °C to 100 °C.

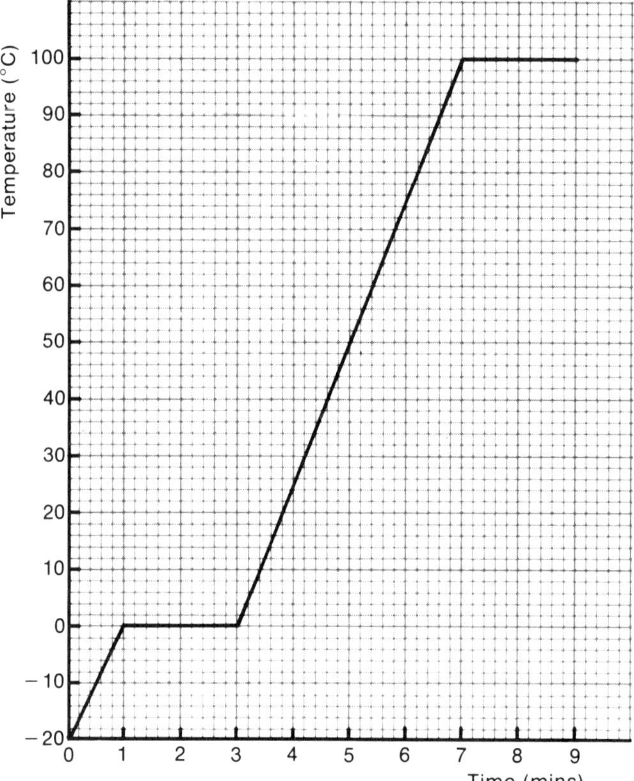

a) What was the temperature after 4 minutes of heating? (1)

b) How long did it take to reach a temperature of 70 °C? (1)

c) How long did the water take to melt? (1)

d) What is happening to the water after 7 minutes? (1)

6 A woman buys a packet of Epsom salts from the chemist. On the side of the packet she notices the following information.

Contents: magnesium sulphate $MgSO_4 \cdot 7H_2O$

Net mass: 492 g

Calculate how many grams of water the woman carried home with her. (2)

7 From the following graphs (**A** to **E**), choose the correct letter for questions (a) to (c).

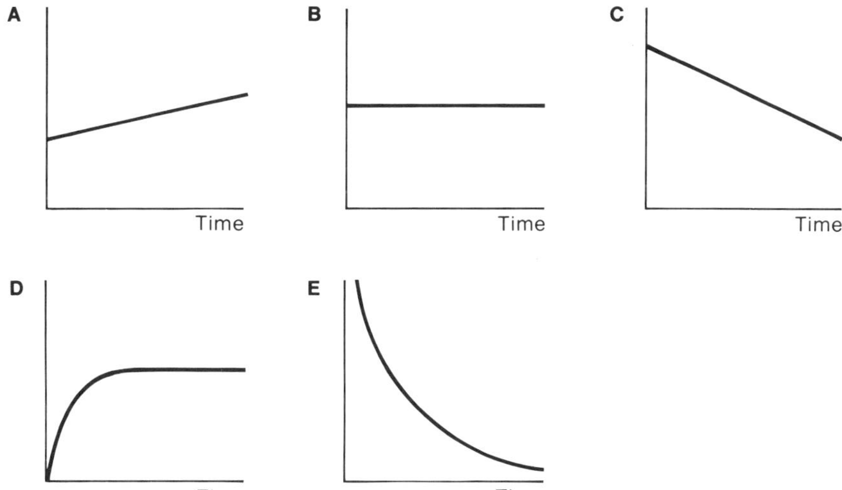

Which of the above sketched graphs best describes

a) the mass of manganese(IV) oxide present during the preparation of oxygen from hydrogen peroxide (1)

b) the rate of production of oxygen (as measured by volume) in the same reaction (1)

c) the count rate for the decay of a radioactive isotope? (1)

8 **Definition**. One mole of solute dissolved in 1000 cm^3 of solution makes a one molar solution.

How many moles of solute are there in

a) 1000 cm^3 of a 0.5 molar solution (1)

b) 200 cm^3 of a 1 molar solution (1)

c) 250 cm^3 of a 0.25 molar solution? (1)

9 One litre of sodium hydroxide solution was made by dissolving 4 g of sodium hydroxide in deionised water.

a) What is the molar concentration (molarity) of this solution? (1)

b) 10 cm^3 of the solution was pipetted into a beaker. How many grams of sodium hydroxide would there be in this sample? (1)

c) What would be the molar concentration (molarity) of the solution in the beaker? (1)

10 For each of the solutions detailed below, work out the molar concentration (molarity).

a) 0.6 moles of sodium nitrate ($NaNO_3$) in 500 cm^3 water (2)

b) 2 moles of copper(II) sulphate ($CuSO_4$) in 100 cm^3 water (2)

11 Which one of the following chemicals contains over 50% by mass of water?

A **$CaSO_4 \cdot 2H_2O$** B **$Na_2S_2O_3 \cdot 5H_2O$** C **$FeCl_3 \cdot 6H_2O$**

D **$MgCl_2 \cdot 6H_2O$** E **$ZnSO_4 \cdot 7H_2O$** (3)

TEST 23
Rates of Reaction

1 Select the missing words from the list below.

chemically decrease enzyme increase large physically small specific

Copy, and complete, the following paragraph on catalysts.

A catalyst is usually used to ____________ the rate of a chemical reaction. It is used in ____________ amounts. Catalysts are unchanged ____________ during a reaction but may be changed ____________. Catalysts are often ____________ to a particular reaction. An ____________ is a natural catalyst. (3) ($\frac{1}{2}$ each)

2 Look at the five test tubes, **A** to **E**, shown below. All contain a small piece of magnesium ribbon (the same size in each tube), in excess dilute hydrochloric acid, and produce hydrogen gas.

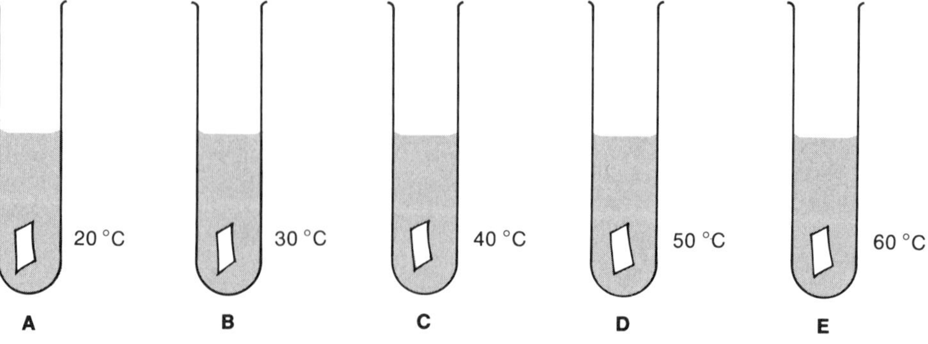

a) In which test tube will the magnesium ribbon be used up most quickly? (1)

b) In which test tube will the slowest reaction take place? (1)

c) If all the reactions were started at the same time, which will be the second test tube to stop producing hydrogen? (1)

3 Why do reactions between gases usually go faster at higher pressures? (1)

4

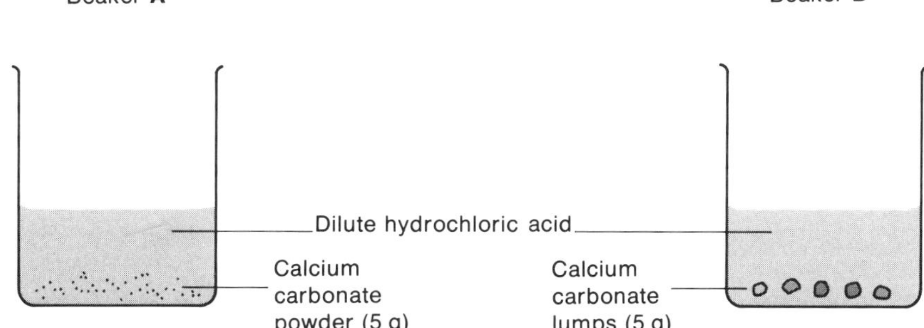

A pupil set up two beakers, as shown above, side by side in the laboratory.

a) In which beaker would the faster reaction take place? (1)

b) Explain your answer to (a) above. (1)

c) Why is it important to use excess dilute hydrochloric acid? (1)

d) *Name* the gas given off from both beakers. (1)

5 **Matching pairs question**. For questions (a) to (c), the group of questions has a set of four alternative answers, **A**, **B**, **C** and **D**. Each letter may be used *once*, *more than once* or *not at all*. Show your choice by writing **A**, **B**, **C** or **D** opposite the question number on your answer sheet. From the list

A 2 g of zinc powder with excess dilute hydrochloric acid at 20 °C

B 2 g of zinc powder with excess dilute hydrochloric acid at 40 °C

C 2 g of zinc granules with excess dilute hydrochloric acid at 20 °C

D 2 g of zinc granules with excess dilute hydrochloric acid at 50 °C

choose which set of conditions would

a) give the slowest rate of production of hydrogen (1)

b) give off hydrogen most quickly (1)

c) produce hydrogen for the longest length of time. (1)

6 A solid may be represented by six cubes (seen from above) arranged in five possible ways, as shown in diagrams **A** to **E** below.

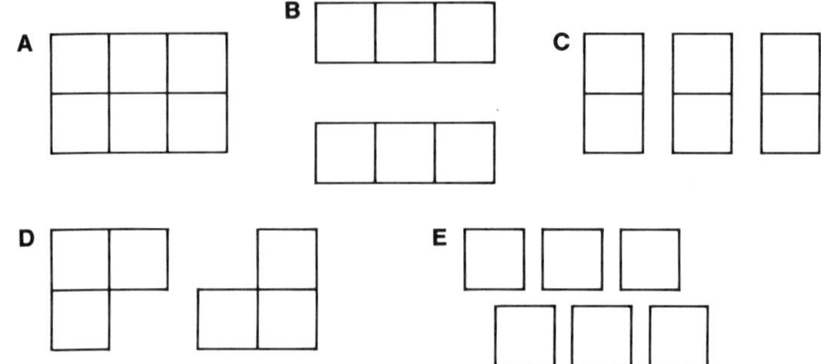

From the diagrams **A** to **E** select

a) the way which would result in the fastest reaction (1)

b) the way which would result in the slowest reaction (1)

c) the two ways which would react at the same speed. (1)

d) Explain your answers to (a) and (b) above. (1)

e) Explain your answer to (c) above. (1)

7 The Haber process for the manufacture of ammonia may be represented by the following equation

$$N_2(g) + 3H_2(g) \rightleftharpoons 2NH_3(g) \qquad \Delta H \text{ is negative}$$

a) What does the sign '$\rightleftharpoons$' mean? (1)

b) What does 'ΔH is negative' mean? (1)

c) Ideally this process should be operated at a very low temperature. Why, in practice, is a temperature of around 500 °C chosen, in preference to the ideal figure? (2)

Questions 8, 9 and 10 concern the reaction to produce oxygen from alkaline hydrogen peroxide, using a suitable catalyst. The graph shows the production of oxygen at room temperature (curve **A**).

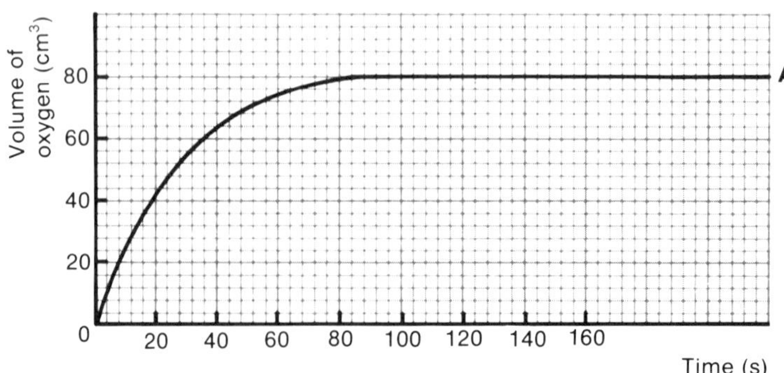

8 Name the catalyst normally used in this experiment. (1)

9 When was the reaction going the fastest?

A **during the first 20 s** B **between 20 and 40 s**

C **between 40 and 60 s** D **between 60 and 80 s**

E **after 80 s** (1)

10 Sketch the graph on the opposite page on your answer paper, and add a curve **B** for oxygen production using a poorer catalyst with exactly the same amount of hydrogen peroxide. (1)

11 At any one temperature, it may be found that the rate of a reaction is directly proportional to the concentration of the reacting chemicals. One such reaction is that between nitric acid and sodium thiosulphate solution.

Sodium thiosulphate + nitric acid → sodium nitrate + water + sulphur + sulphur dioxide

You are asked to study this reaction, at normal laboratory temperature, with the nitric acid concentration fixed throughout.

a) If you halve the concentration of the sodium thiosulphate solution in successive experiments, what would you notice about the rate of production of the sulphur? (1)

b) How can the result in (a) above be explained in terms of molecules and their behaviour? (1)

c) Suggest a simple way of measuring the rate of sulphur produced in this experiment. (2)

TEST 24
Energy

1 In questions (a) and (b), choose the correct answers from the list.

A **fossil fuels** B **wood fuel** C **nuclear fuel**

D **wind** E **waves** F **sun**

a) Give *two* sources of energy which are renewable. (2)

b) Give *two* sources of energy which are non-renewable. (2)

2 Copy and complete the following passage.

A chemical reaction can be detected by a rise or fall in ___________, which can be measured directly with a ___________. This is due to ___________ energy being given out to or taken in from the surroundings. (3)

3 Zinc reacts with dilute sulphuric acid to produce hydrogen. What effect does an increase in energy (in the form of heat) have upon

a) the rate at which hydrogen is produced (1)

b) the maximum volume of hydrogen that it is possible to produce? (1)

4 What kind of reaction is defined as giving out more heat energy than is needed to start the reaction? (1)

5 The graph below shows a solid being heated from room temperature. The source of heat was never removed during the experiment.

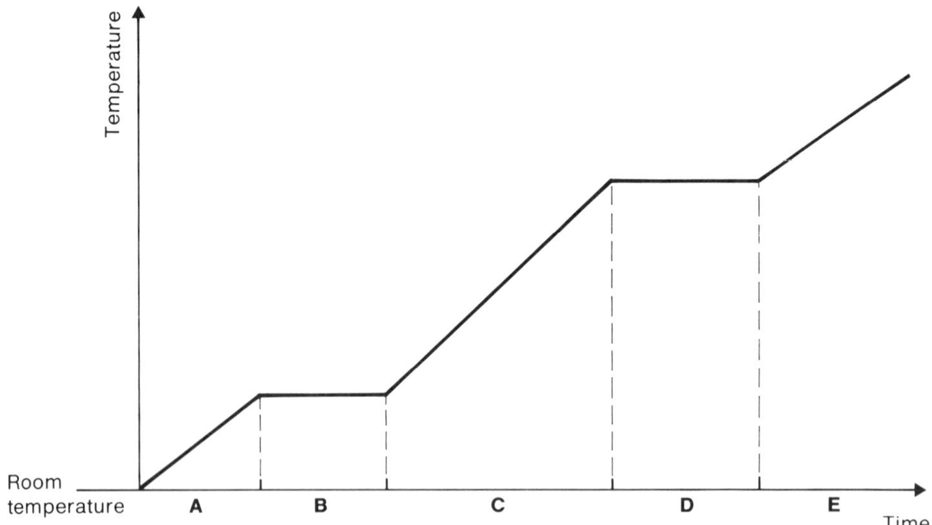

a) During which time period (**A** to **E**) was the substance a liquid? (1)

b) During which time period was the substance melting? (1)

c) Which state of matter is the substance in during time period **E**? (1)

d) Energy is being supplied to the substance during time period **B**, yet the temperature is not rising. What is the energy being used for? (2)

6 Which of the following reactions is endothermic?

A **combustion of fuels**

B **hydration of copper(II) sulphate (anhydrous)**

C **neutralisation**

D **photosynthesis**

E **respiration** (1)

7 Look at the diagram below. At room temperature the gas syringe reads 50 cm^3.

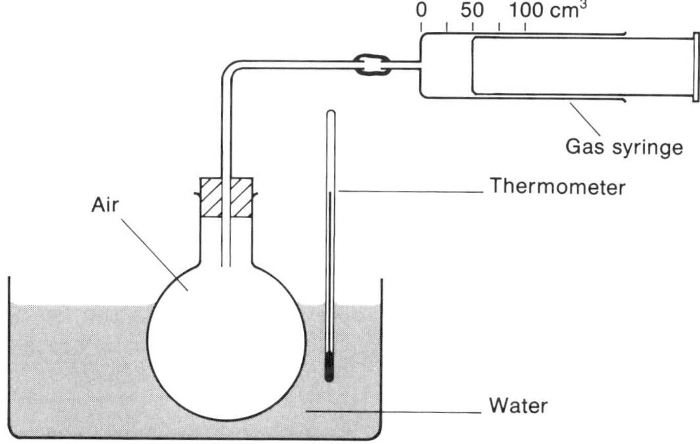

For parts (a) and (b) below, choose the correct response from this list.

A **The air molecules get bigger.**

B **The air molecules get faster.**

C **The air contracts.**

D **The ice molecules lose energy.**

Which of these explanations best fits the following observations?

a) Some ice was added to the water. The syringe read 45 cm^3. (1)

b) The water was warmed to 50 °C. The syringe read 55 cm^3. (1)

8 Which of the following is true for reactions giving out heat?

A ΔH = **the temperature rise**

B ΔH = **the temperature fall**

C ΔH = **positive**

D ΔH = **zero**

E ΔH = **negative** (1)

9

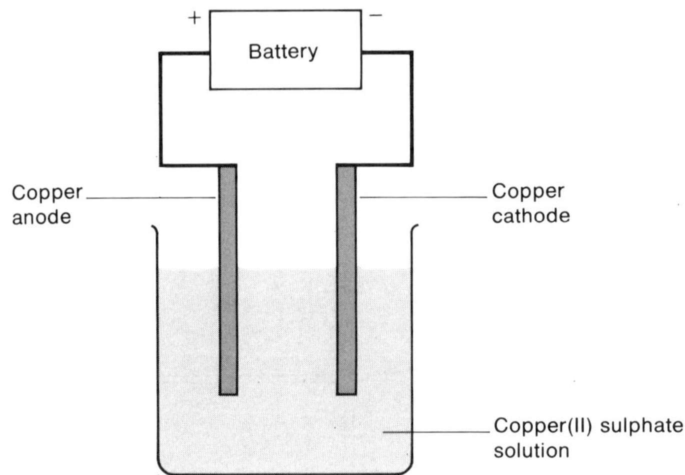

a) Which electrode will gain mass during the experiment? (1)

b) Copy the *one* graph below which shows the mass of the electrode chosen in part (a) if a constant current (i.e. constant electrical energy) is applied. (1)

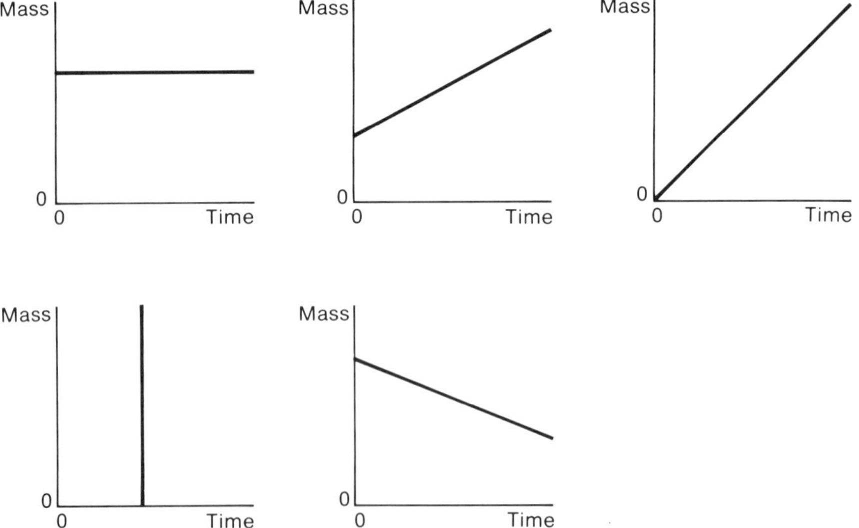

c) On to this graph add a second line which shows the same experiment repeated again (from time = 0) except at a higher current. (1)

10 What are the main forms of energy produced from

a) a fuel cell (1)

b) combustion of methane? (2)

11

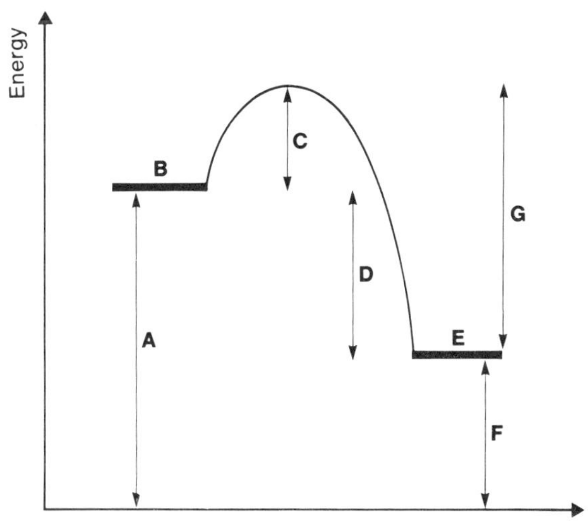

On the diagram above, which letter (**A** to **G**) represents

a) the reactants (1)

b) the products (1)

c) the activation energy (1)

d) ΔH for the reaction? (1)

e) What effect would a catalyst have on the activation energy? (1)

TEST 25
General Gas Properties and Kinetic Theory

1 Name the three physical states of matter. (1)

2 **Matching pairs question**. For questions (a) to (c), there are four alternative answers, **A** to **D**. Each letter may be used *once, more than once* or *not at all*. Show your choice by writing the letter opposite the question number on your answer sheet.

From the diagrams **A** to **D** below

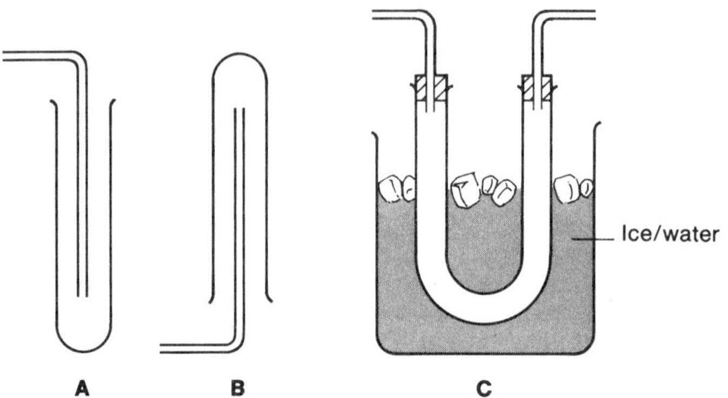

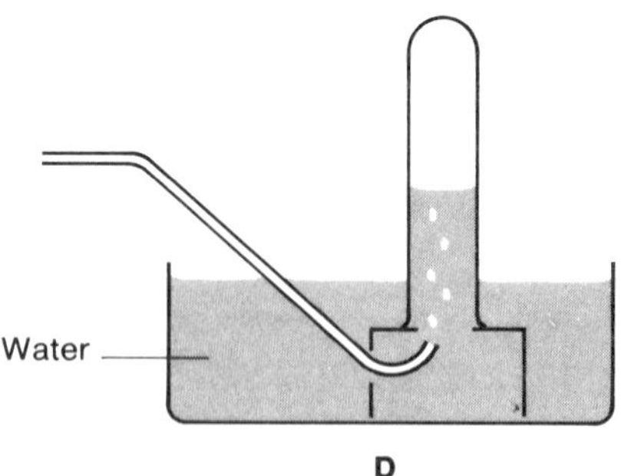

choose the apparatus best suited for the collection of

a) ammonia gas (1)

b) dry hydrogen (1)

c) dry carbon dioxide. (1)

3 **Matching pairs question**. For questions (a) to (i) there are six alternative answers, **A** to **F**. Each letter may be used *once, more than once* or *not at all.* Show your choice by writing the letter opposite the question number on your answer sheet.

From the list

A	**ammonia**	B	**carbon dioxide**	C	**hydrogen**
D	**nitrogen**	E	**oxygen**	F	**sulphur dioxide**

choose the gas which

a) is used to make the fizz in fizzy drinks (1)

b) is an unreactive gas at room temperature and pressure (1)

c) is the most soluble in water (1)

d) is called 'dry ice' in the solid state (1)

e) is a diatomic gas not found in the atmosphere (1)

f) dissolves in water giving an alkaline solution (1)

g) contains a triple bond within its molecule (1)

h) is needed to support life (1)

i) contains the largest number of atoms within one molecule of the gas. (1)

4 In an experiment to study the diffusion of gases, concentrated ammonia solution and concentrated hydrochloric acid were placed in the apparatus shown below.

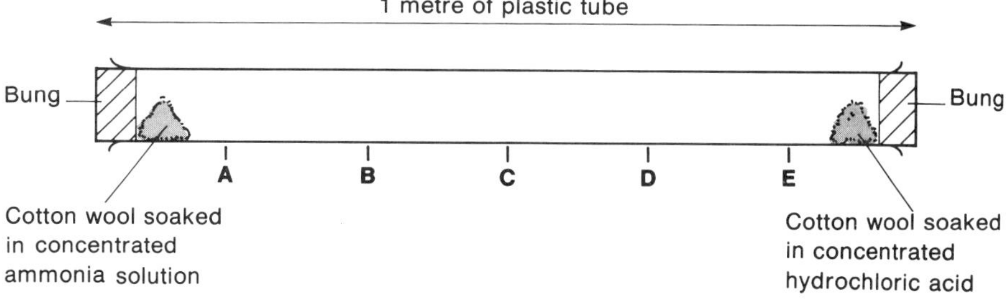

a) Which of the letters **A**, **B**, **C**, **D** or **E** best represents the point at which the gases would meet? (1)

b) Explain the result in terms of the rate of movement of gas molecules. (2)

c) If the same experiment could be carried out with a vacuum in the tube, how, if at all, would this change the result? (2)

5

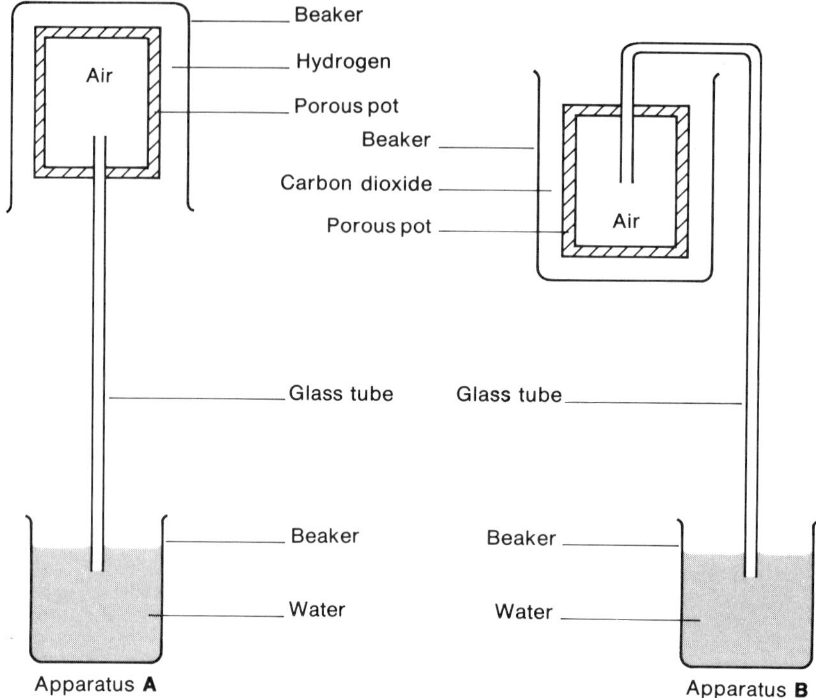

The above sets of apparatus have been set up to study gaseous diffusion.

a) Why is it necessary to use two different designs of apparatus? (1)

b) State, and explain, what will be observed in apparatus **A**. (2)

c) State, and explain, what will be observed in apparatus **B**. (2)

6 a) Explain why hydrogen is preferred to helium in meteorological balloons. (1)

b) Explain why helium is preferred to hydrogen in the airships of today. (1)

7 Explain the effect of increasing the pressure separately on a gas and a liquid. (2)

8 Describe briefly what happens to the molecules in water as it is heated from −2 °C to 100 °C. (3)

TEST 26
Chemistry and the Environment

1 The exhaust gases from car engines are a major cause of air pollution.

a) What is the name of the fuel used in most car engines? (1)

b) What two elements are contained in this fuel? (2)

c) If this fuel is burned in an abundance of air, there are two products. Name them. (2)

d) Inside a car engine there is only incomplete combustion, which produces the gas that causes much of the pollution. Name this gas. (1)

2 **Matching pairs question**. For questions (a) to (c) there are five alternative answers, **A** to **E**. Each letter may be used *once, more than once* or *not at all*. Show your choice by writing the letter opposite the number on your answer sheet.

A chlorination

B distillation

C filtration

D flocculation

E neutralisation

Which of the above processes

a) is used to get drinking water from sea water (1)

b) is used to remove suspended solids from sewage water (1)

c) is used to sterilise drinking water before distribution from water treatment works? (1)

3 Sulphur is an impurity in fossil fuels.

a) Name the compound formed from the sulphur as these fuels are burned. (1)

b) What happens when this compound dissolves in rain? (1)

c) Give two problems that are caused by such rain. (2)

4 Why is hydrogen classed as a 'clean fuel'? (3)

5 Gannets are birds which dive from a height in order to catch fish from the sea. Occasionally, oil tankers illegally flush out their tanks at sea causing an oil slick. Explain, with reference to the physical properties of oil *and* water, how these slicks, which are a hazard to gannets, are formed. (2)

6 A farmer had a field in which his crops grew poorly, year after year. Beside the field was a thriving pond. He decided to add fertiliser to his soil and each year the yield from his crops got better. Unfortunately, the pond became stagnant and full of algae. The fish in the pond died. When analysed, the pond water contained some of the fertiliser.

a) Despite careful spreading, how did the fertiliser get into the pond? (1)

b) Why did the algae increase? (1)

c) How did this manage to kill off the fish? (1)

d) A farmer friend does not use fertiliser. He rotates his crops between fields so that every few years, each field will have peas or beans grown in it. Why doesn't he need to use fertiliser? (1)

7 Read the following passage.

The textile mills in the town of Castleford were a source of pollution resulting from detergents. It was once common to see thick foam on the river which was blown, by the wind, all over the town. It was discovered that the problem was due to the detergent being non-biodegradable. Now that new detergents are used, the problem has gone away.

a) What is the detergent used for in the mills? (1)

b) How could the detergent get into the river? (1)

c) What is meant by 'non-biodegradable'? (1)

d) Give one reason why detergents are more useful than soap. (1)

8 The map on the opposite page shows five possible sites for a power station (labelled **A** to **E**). It has not been decided whether to build a coal-fired power station or a nuclear power station.

a) Which site is best suited to a nuclear power station? (1)

b) Give a reason for your answer to (a). (1)

c) Which site is best suited for a coal-fired power station? (1)

d) Give a reason for your answer to (c). (1)

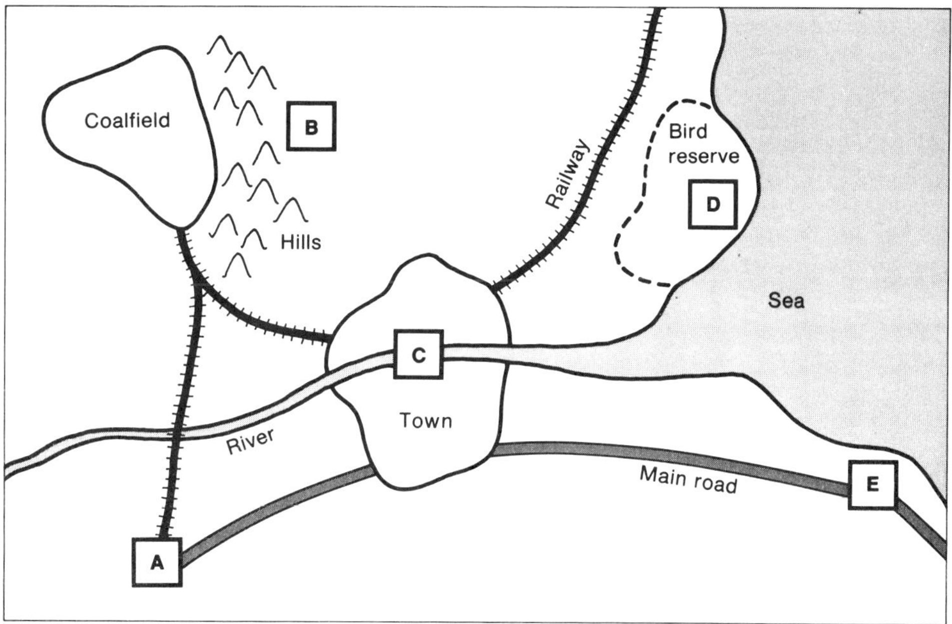

TEST 27
Chemistry, Society and Technology

1 **Matching pairs question**. For questions (a) to (d) there are five possible answers, **A** to **E**. Each letter may be used *once, more than once* or *not at all*. Show your choice by writing the letter opposite the number on your answer sheet.

A **chlorine**

B **copper**

C **iron**

D **silicon**

E **zinc**

Which of these elements

a) is the main constituent of the alloy used in coins (1)

b) is the microelectronics industry founded on (1)

c) is used to make magnets (1)

d) is used as the covering layer in galvanising? (1)

2 From the list below choose the correct response.

A **Lead is a dense metal.**

B **Lead is malleable.**

C **Lead forms coloured compounds.**

D **Lead is toxic.**

E **Lead has a low melting point.**

Which of the above is a reason for

a) lead being used for 'flashings' to stop rain running down chimneys (1)

b) the growing campaign for lead-free petrol (1)

c) the use of lead in the alloy solder? (1)

3 Which acid is used as an electrolyte in car batteries? (1)

4 Which metal's compounds are good for developing bones? (1)

5 Why is calcium carbonate sometimes used by farmers on the soil of their fields? (1)

6 Why does sodium hydrogencarbonate (a compound in baking powder) help cakes to rise when they are baked? (2)

7 An ester commonly used in schools is ethyl ethanoate. Give two uses for esters in general. (2)

8 Rock salt is widely used to de-ice roads in winter.

a) Name the compound contained in rock salt which is responsible for this. (1)

b) What happens to the melting point of ice when this compound dissolves in water? (1)

c) Why does this help to get rid of the ice? (1)

9 Use the following ideas to explain why caves are formed in limestone rocks.

Carbon dioxide is soluble in water.

Calcium carbonate is 'dissolved' by acids. (4)

10 Charcoal is used in air filters to remove some poisonous gases and is also used as a decolourising agent in some reactions. What is this process called?

A **absorption**

B **adsorption**

C **catalysis**

D **diffusion**

E **dissolving** (1)

11 The ingredients on a bottle of lemonade included

a) carbon dioxide

b) glucose

c) citric acid.

For each of the above ingredients, give one reason why it should be in the lemonade. (3)

12 a) Why do aluminium smelting plants often have their own power station? (1)

b) Why are they often built near coal fields? (1)

c) What property of aluminium makes it suitable for

i) aircraft bodies (1)

ii) window frames (1)

iii) electricity power lines? (1)

TEST 28
Chemistry and Industry

1

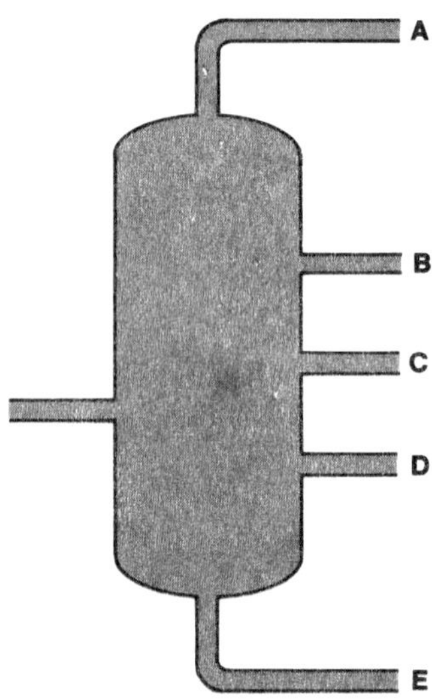

In this representation of the fractionating column in an oil refinery, the position at which the fractions are collected are labelled **A** to **E**.

a) At which position would the gases from petroleum oil be collected? (1)

b) At which position would the bitumen be collected? (1)

c) Name one alkane that would be in the petroleum gases. (1)

d) Give two uses for the bitumen. (2)

2 When coal is heated in the absence of air there are four main products. Coal gas and coal tar are two of them.

a) The third is a pungent-smelling gas which dissolves in water to produce an alkali. Name it. (1)

b) The fourth is the residue from the coal. Name it. (1)

c) What is the main difference between using this residue as a fuel and using coal? (1)

d) Coal gas is no longer supplied to homes. What has replaced it? (1)

3 In making margarine, unsaturated oils are converted into saturated fats by the process known as hydrogenation.

a) What gas is used for the hydrogenation? (1)

b) What effect does this have on the melting point of the oils? (1)

4 Table salt is manufactured from salt-bearing rocks in the ground, called rock salt.

a) What important use does rock salt have in winter? (1)

b) How are the impurities removed from the rock salt? (2)

c) At the end of the refining process, the salt obtained is in a solution. How are the crystals of salt obtained? (1)

d) What is the chemical name for the main ingredient in table salt? (1)

5 Ammonium nitrate and ammonium sulphate are two chemicals used for fertilisers.

a) From the list below, choose a useful element that they both put back into the soil.

A **ammonia**

B **hydrogen**

C **oxygen**

D **nitrogen**

E **sulphur** (1)

What acid is needed to make

b) ammonium nitrate from aqueous ammonia (1)

c) ammonium sulphate from aqueous ammonia? (1)

d) Choose *two* other elements from the list below which are major parts of commercial fertilisers.

A **aluminium** B **carbon** C **potassium** D **chlorine**

E **phosphorus** F **iodine** G **sodium** H **iron** (2)

6 The metal aluminium is extracted from its ore by electrolysis.

a) What is the name of the main aluminium ore? (1)

b) Why is the solid ore not electrolysed? (1)

c) Why is the ore dissolved in a solvent rather than being electrolysed in the molten state? (1)

d) Name the electrode at which the aluminium is collected. (1)

7

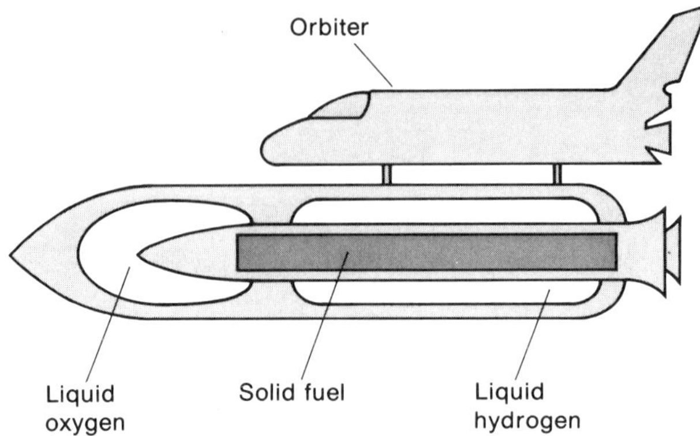

The space shuttle above uses two main fuels to get it into orbit. The large central tank contains hydrogen and oxygen in liquid form. The side rockets contain a solid fuel. The solid fuel is a modification of the thermit reaction between aluminium and iron(III) oxide.

a) What chemical is made by burning hydrogen and oxygen? (1)

b) Write a balanced symbol equation for the thermit reaction. (2)

c) The thermit reaction is a redox reaction. What is oxidised in the reaction? (1)

d) Give another use for the thermit reaction. (1)

TEST 29
Isotopes, Radioactivity and Nuclear Energy

1 The diagrams below show the structures of three possible isotopes of hydrogen.

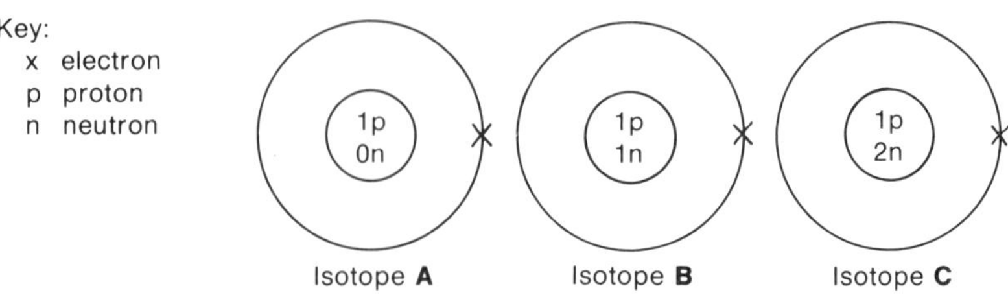

a) Define 'isotopes'. (1)

b) Name isotope **B**. (1)

c) Isotopes can be written in a short-hand way, for example:

Mass number ⟶ A
Atomic number ⟶ Z
$^{A}_{Z}X$ ⟵ Symbol of the element

Write isotope **C** in this way. (1)

2 Copy and complete the following table.

Isotope	*Number of protons*	*Number of neutrons*	*Number of electrons*
$^{4}_{2}He$			
$^{12}_{6}C$			
$^{41}_{19}K$			
$^{238}_{92}U$			

(1)
(1)
(1)
(1)

3 The accurate relative atomic mass of copper is 63.546. It exists as two isotopes with mass numbers 63 and 65. Which isotope is the more abundant? (1)

4 a) Give the names or symbols of the two particles which may be lost from radioactive isotopes. (1)

b) Give the name or symbol of the energy often accompanying loss of particles from the nuclei of radioactive isotopes. (1)

5 A radioactive isotope gives off three types of radiation. A protective screen is placed round it, consisting of paper, 6 mm thick aluminium and finally thick concrete, as shown below (in cross section).

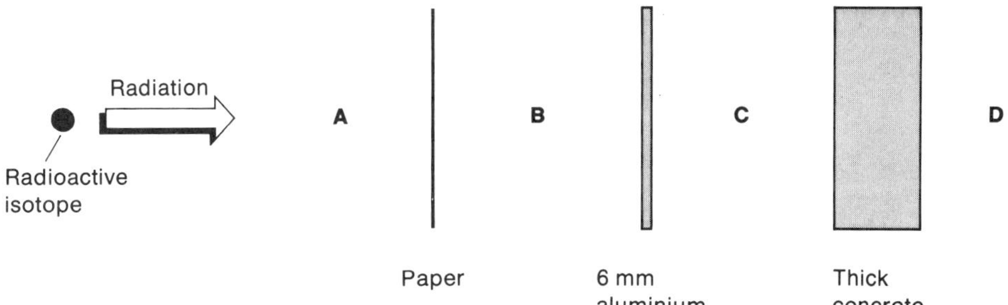

Give the names or formulae of the types of radiation found, if any, at points **A**, **B**, **C** and **D**. (2) ($\frac{1}{2}$ mark each)

6 The half-life of iodine-131 is 8 days. What do we mean by half-life? (1)

7 The half-life of carbon-14 is 5730 years.

a) If we have 100 g of this isotope, how many grams of it will be radioactive after

i) 5730 years (1)

ii) 11 460 years? (1)

b) What use can be made of carbon-14 and its long half-life? (1)

8 **Matching pairs question**. For questions (a) to (e), there are four alternative answers, **A** to **D**. Each letter may be used *once, more than once* or *not at all*. Show your choice by writing the letter opposite the number on your answer sheet. From the list

A $^{235}_{92}U$ **B** $^{60}_{27}Co$ **C** $^{131}_{53}I$ **D** $^{239}_{94}Pu$

choose the isotope which is used

a) to find the efficiency of the human kidney (1)

b) in fast breeder reactors (1)

c) in the treatment of some types of cancer (1)

d) in the early types of nuclear power station (1)

e) in a nuclear battery in a heart pacemaker. (1)

9 Some water from a nuclear power station has become slightly radioactive due to an accident. Why has the water to be stored rather than discharged into the sea? (1)

10 Why do both isotopes of the element chlorine have the same chemical properties? (2)

11 Why do radioactive isotopes eject particles from their nuclei? (1)

12 Sodium and magnesium are next to one another in the periodic table. Their atomic numbers are 11 and 12, respectively. Why is it very unlikely that a planet in another galaxy will contain an element which can be placed between them? (1)

13 Uranium-235 and carbon-14 are naturally radioactive. What does this mean? (1)

14 Plutonium-239 is artificially radioactive. What does this mean? (1)

15 The recently discovered wonder element batlium has been found to have two isotopes, **A** and **B**.

Isotope	*Half-life*
A	60 minutes
B	6000 years

These isotopes are both dangerously radioactive. Explain why the disposal of isotope **B** is so much more of a problem than is that for isotope **A**. (2)

TEST 30
Identification Tests

1 A student finds that the melting point of an unknown solid is between 78 °C and 80 °C. The teacher says that the melting point should be 81 °C.

a) Is the student's unknown solid pure or impure? (1)

b) Explain your answer to (a) above. (2)

2 **Matching pairs question.** For questions (a) to (i) there are five alternative answers, **A** to **E**. Each letter may be used *once, more than once* or *not at all.* Show your choice by writing your answer opposite the question number on your answer sheet. From the list

A **hydrogen**

B **oxygen**

C **ammonia**

D **ethene**

E **methane**

choose the gas which you would normally expect to

a) relight a glowing splint (1)

b) 'pop' in contact with a lighted splint (1)

c) decolourise bromine water (1)

d) be very soluble in water (1)

e) support combustion (1)

f) be a saturated hydrocarbon (1)

g) give an alkaline reaction with moist universal indicator paper (1)

h) be an element, and a good reducing agent (1)

i) be oxidised to nitrogen and water. (1)

3 Copy and complete the following.

Limewater is a solution of ____________ ____________. When carbon dioxide is bubbled through, the solution goes ____________. The new chemical produced is ____________ ____________. When excess carbon dioxide is bubbled through, the solution now contains ____________ ____________ and is temporarily hard water. (4)

4 A student is given two white, powdered solids. The teacher says that one is zinc oxide, the other is sodium hydrogencarbonate. They are labelled **X** and **Y**. How could the student identify them by the effect of heat alone? (2)

5 How would you test to show that a solution contained water? (1)

6 How would you test to show that a liquid was *pure* water? (1)

7 You are given a sample of hard water. It is *either* temporarily hard *or* permanently hard. Describe, briefly, how you would go about finding which it was. (3)

8 You are given a black solid. The teacher tells you that it is *either* copper(II) oxide *or* manganese(IV) oxide. Describe a test or tests you would perform in order to identify which chemical it was. (2)

9 You are given two yellow powders and told that one is lead(II) oxide whilst the other is sulphur. Describe how, with one simple test, you could tell the difference between them. (2)

10 Three bottles have lost their labels. They are known to be the following.

dilute ammonia solution

dilute hydrochloric acid

dilute sulphuric acid

A student is now given six test tubes, a bottle of universal indicator solution, and a bottle of barium chloride solution. Outline how, using only these things, the bottles could be identified and hence relabelled. (3)

REACTION SCHEMES

REACTION SCHEME 1

Acids, Bases and Salts

Part 1

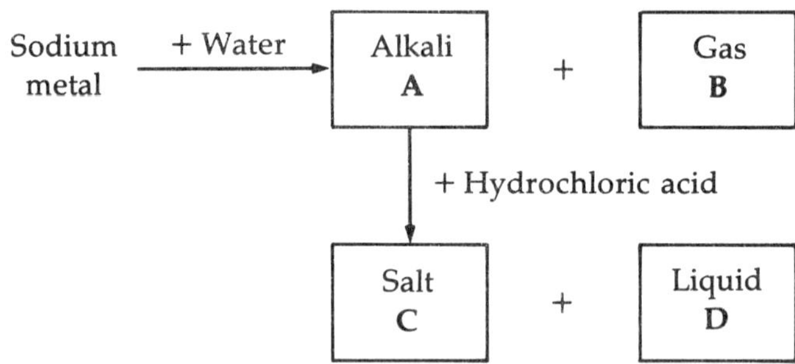

Questions Give the name *or* formula of

1 alkali **A**

2 gas **B**

3 salt **C**

4 liquid **D**.

Part 2

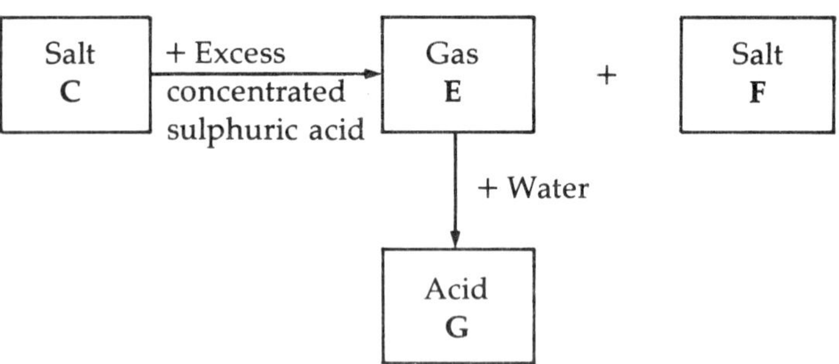

Questions Give the name *or* formula of

5 gas **E**

6 salt **F**

7 acid **G**

8 the element needed to convert hydrogen into gas **E**.

Part 3

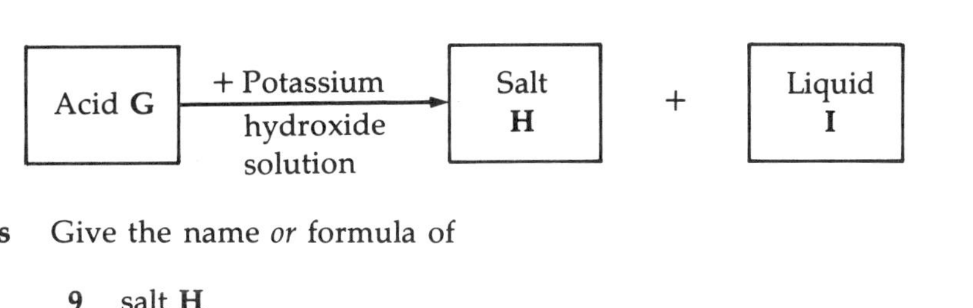

Questions Give the name *or* formula of

9 salt **H**

10 liquid **I**.

REACTION SCHEME 2
Sodium Hydroxide Solution

Part 1

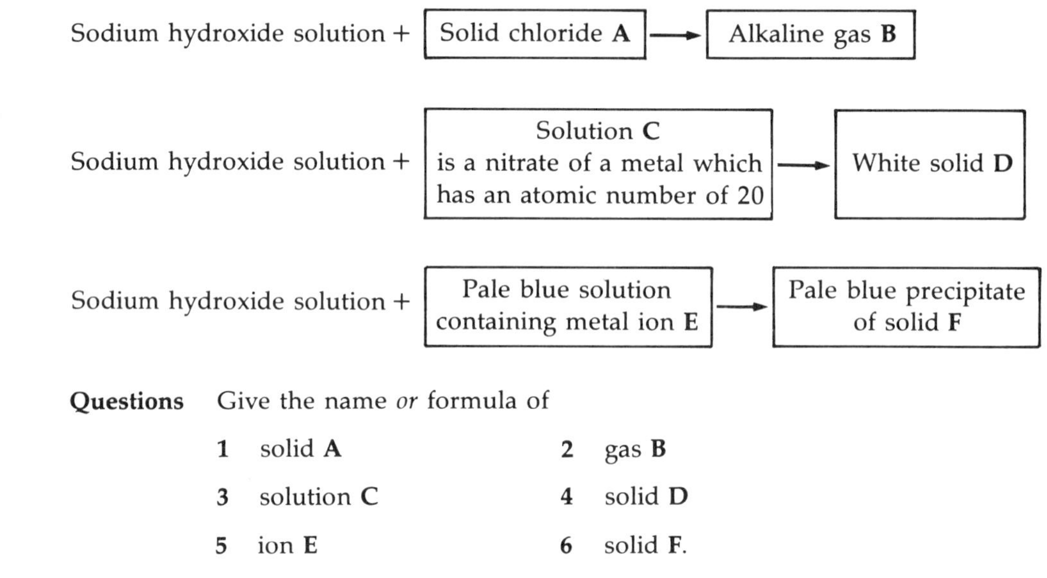

Questions Give the name *or* formula of

1 solid **A** **2** gas **B**

3 solution **C** **4** solid **D**

5 ion **E** **6** solid **F**.

Part 2

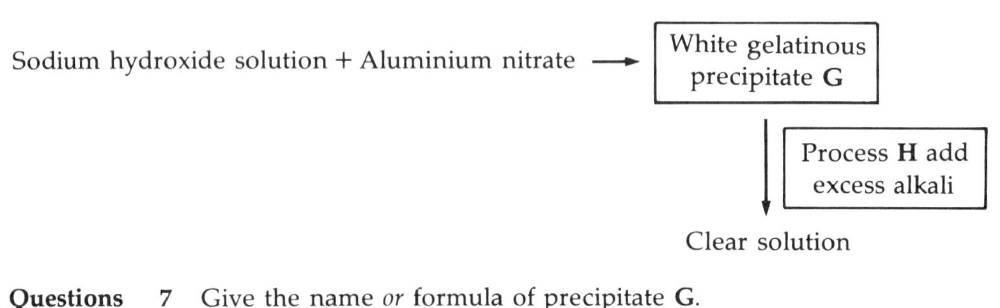

Questions **7** Give the name *or* formula of precipitate **G**.

8 What property of **G** is shown by process **H**?

Part 3

Sodium hydroxide solution + Glyceryl stearate ⟶ Glycerol + Compound **I**

Questions **9** Give the chemical name for compound **I**.

10 What is the common household name for **I**?

REACTION SCHEME 3
The Gases in Air

Part 1

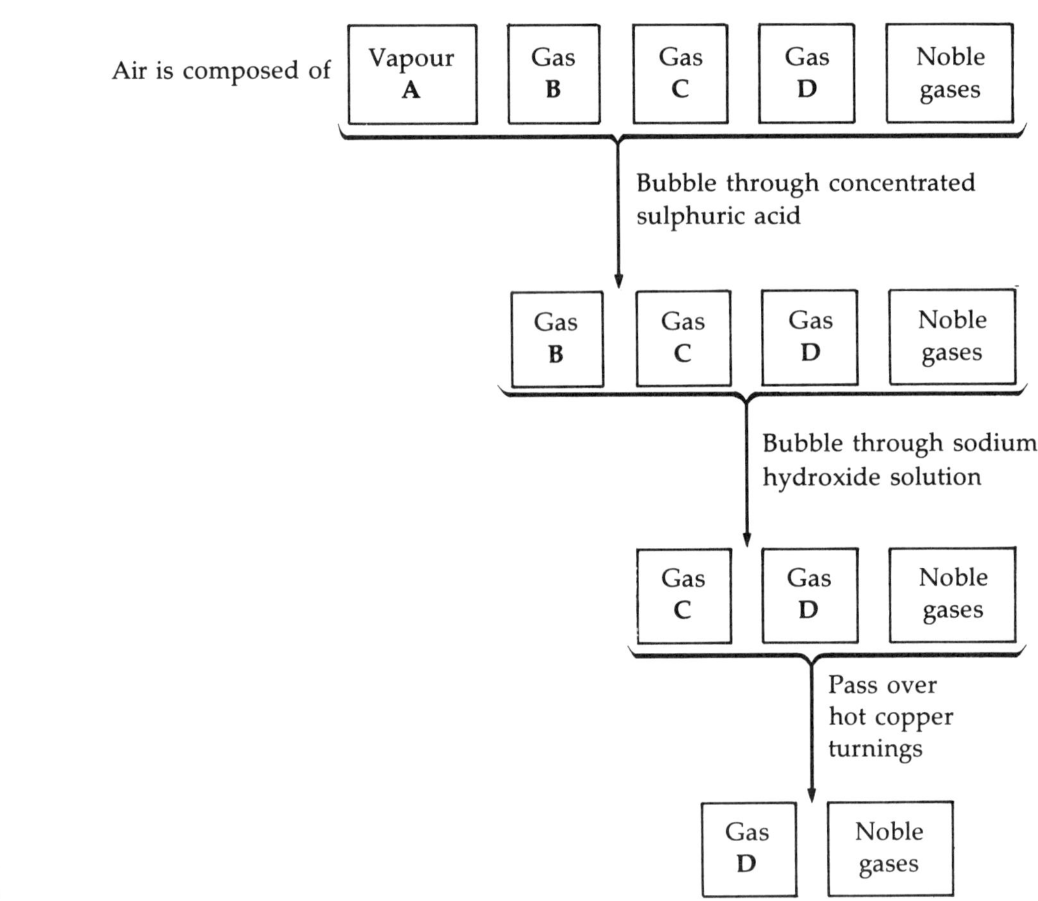

Questions Give the name *or* formula of

1 vapour **A**

2 gas **B**

3 gas **C**

4 gas **D**

5 one noble gas.

Part 2

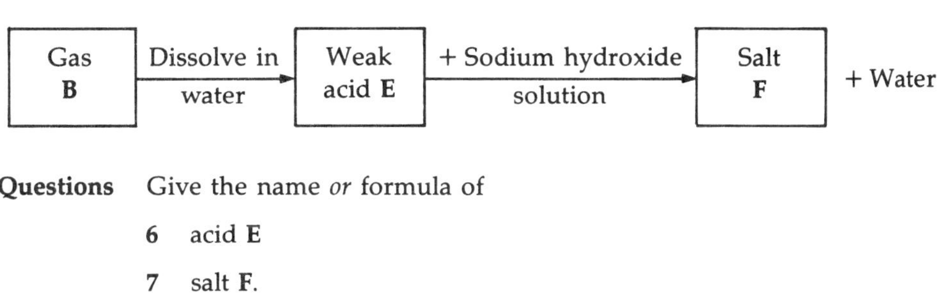

Questions Give the name *or* formula of

6 acid **E**

7 salt **F**.

Part 3

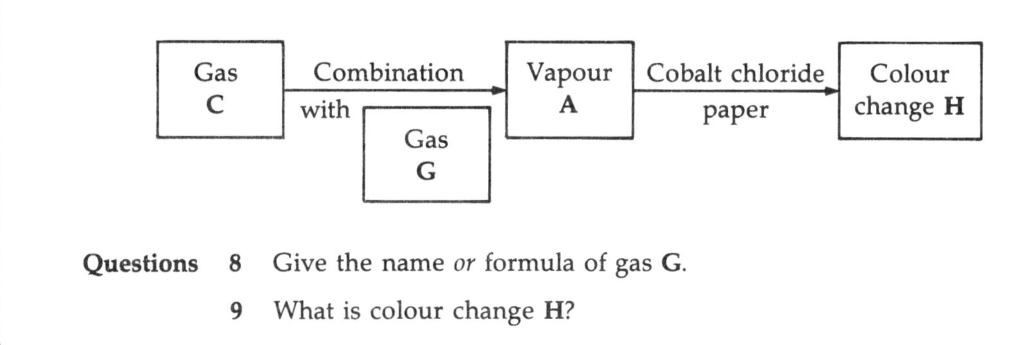

Questions **8** Give the name *or* formula of gas **G**.

9 What is colour change **H**?

Part 4

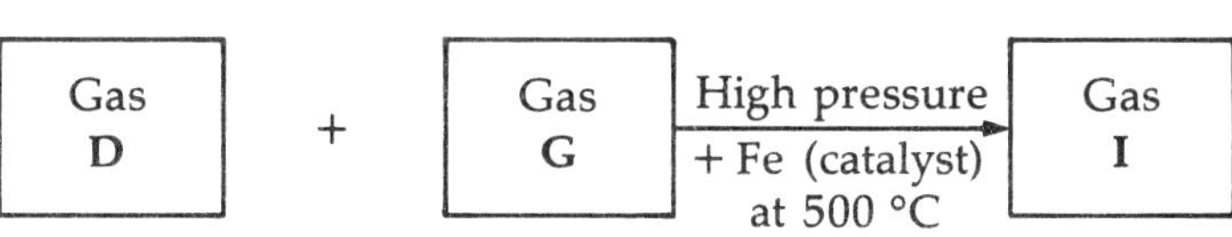

Question **10** Give the name *or* formula of Gas **I**.

REACTION SCHEME 4
Oxides

Part 1

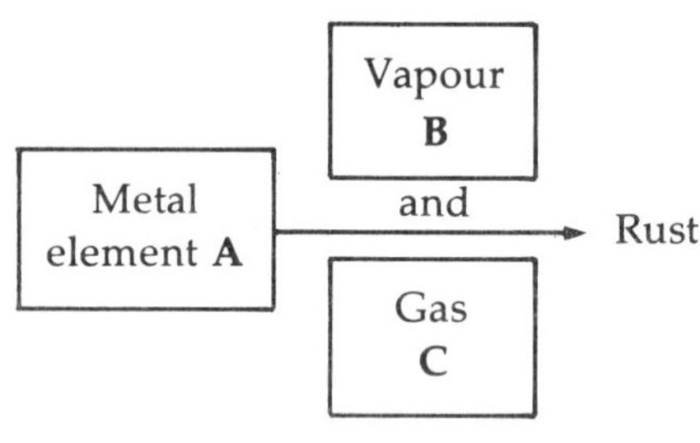

Questions Give the name *or* formula of

1 metal element **A** **2** vapour **B**

3 gas **C**.

Part 2

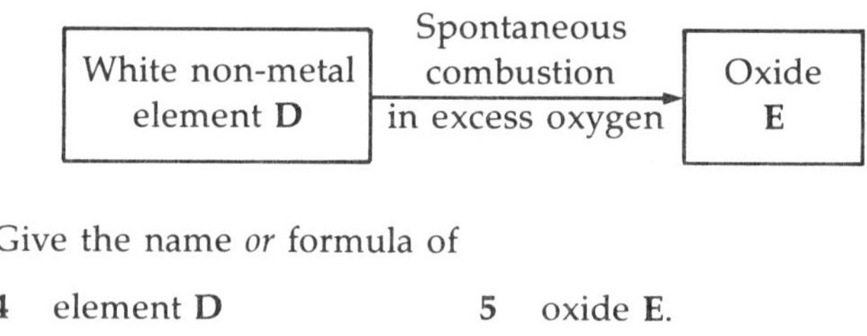

Questions Give the name *or* formula of

4 element **D** **5** oxide **E**.

Part 3

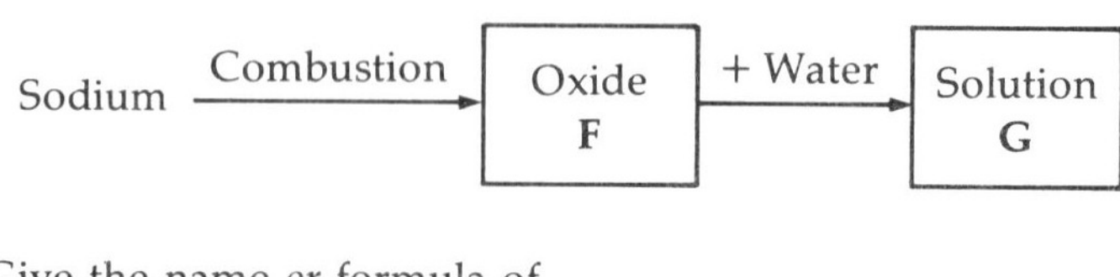

Questions Give the name *or* formula of

6 oxide **F** **7** solution **G**.

Part 4

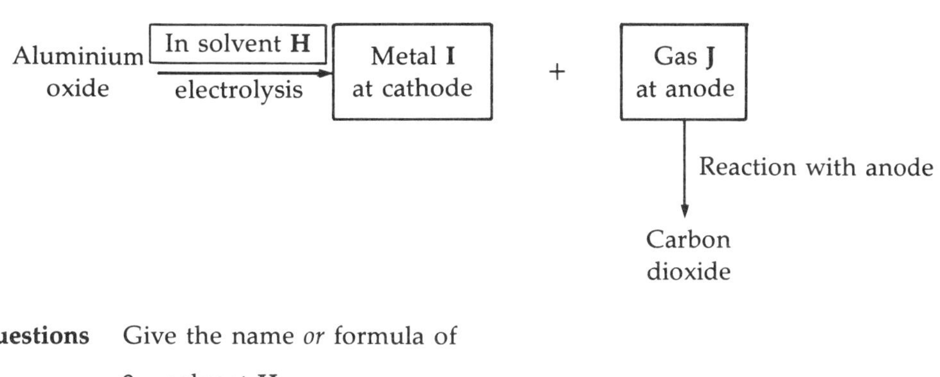

Questions Give the name *or* formula of

8 solvent **H**

9 metal **I**

10 gas **J**.

REACTION SCHEME 5
Hydrogen and Water

Part 1

Alkali metal **A** + Water --Vigorous reaction / violet flame--> Solution **B** + Gas **C**

Questions Give the name *or* formula of

1 metal **A**

2 solution **B**

3 gas **C**.

Part 2

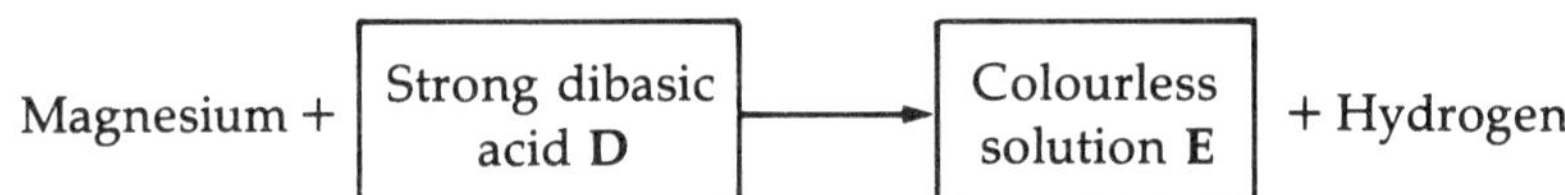

Questions Give the name *or* formula of

4 acid **D**

5 solution **E**.

Part 3

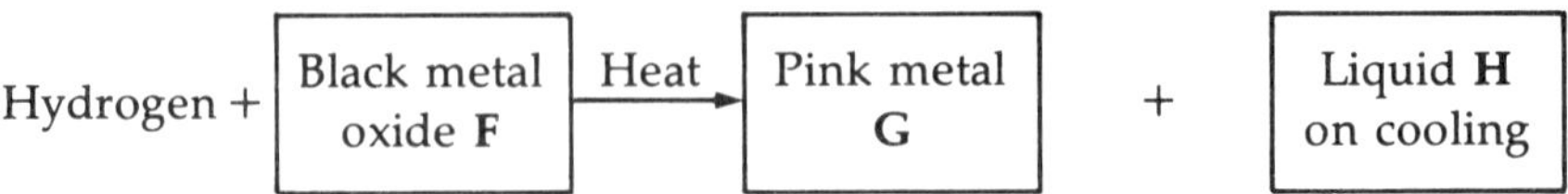

Questions Give the name *or* formula of

6 oxide **F**

7 metal **G**

8 liquid **H**.

Part 4

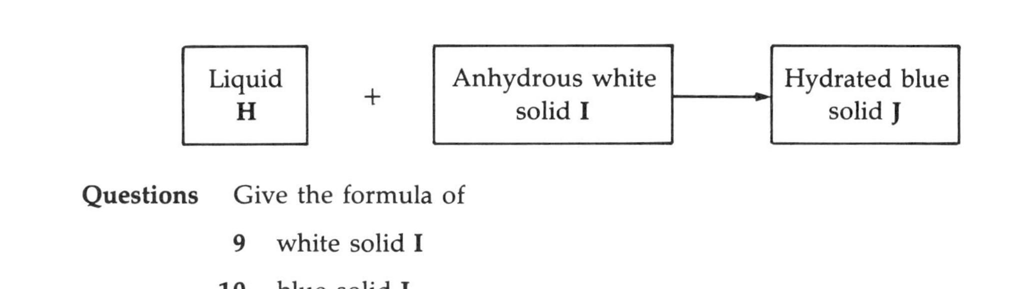

Questions Give the formula of

9 white solid **I**

10 blue solid **J**.

REACTION SCHEME 6
Inorganic Carbon Chemistry

Part 1

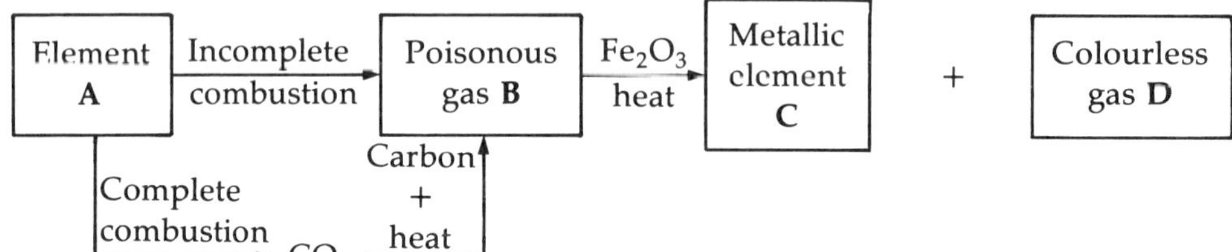

Questions Give the name *or* formula of

1 element **A**

2 poisonous gas **B**

3 metallic element **C**

4 colourless gas **D**.

Part 2

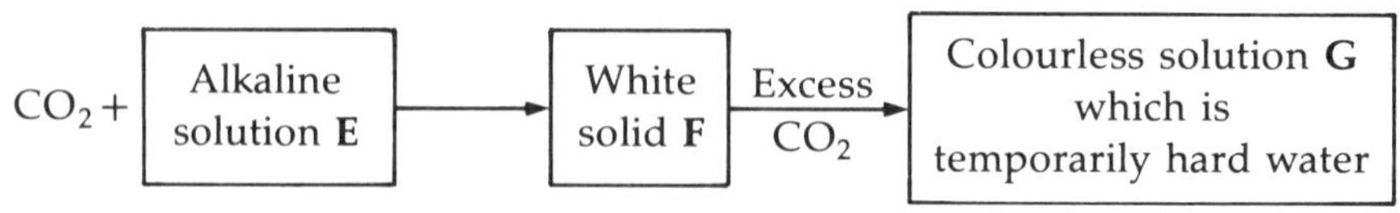

Questions Give the name *or* formula of

5 alkaline solution **E**

6 white solid **F**

7 colourless solution **G**.

Part 3

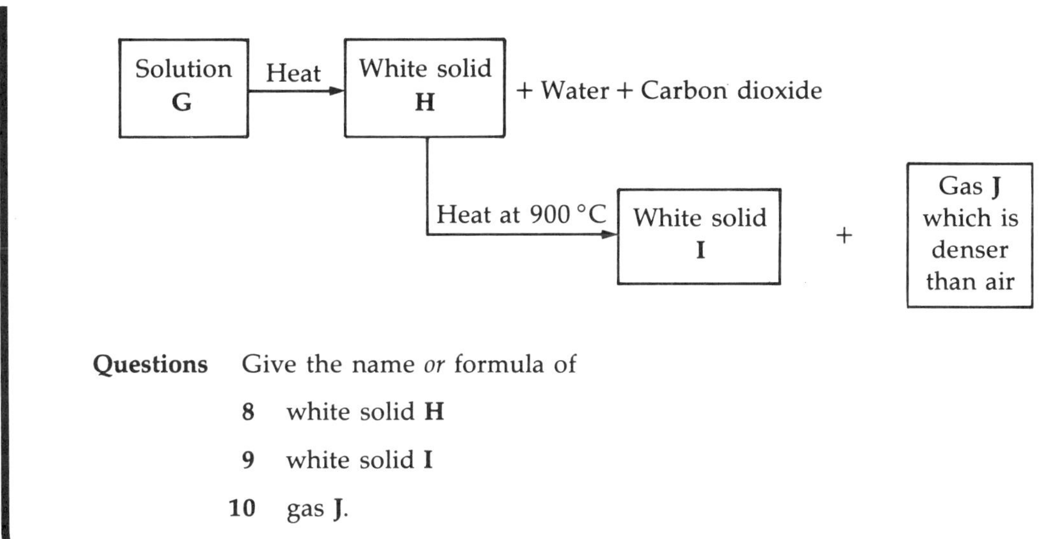

Questions Give the name *or* formula of

8 white solid **H**

9 white solid **I**

10 gas **J**.

REACTION SCHEME 7
Calcium Carbonate

Part 1

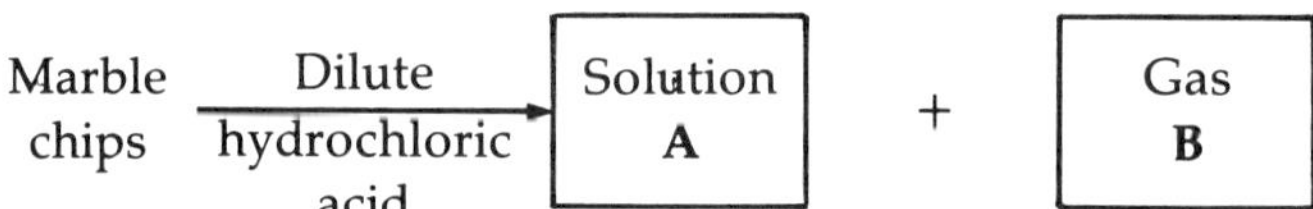

Questions Give the name *or* formula of

1 solution **A**

2 gas **B**.

Part 2 (Natural occurrence)

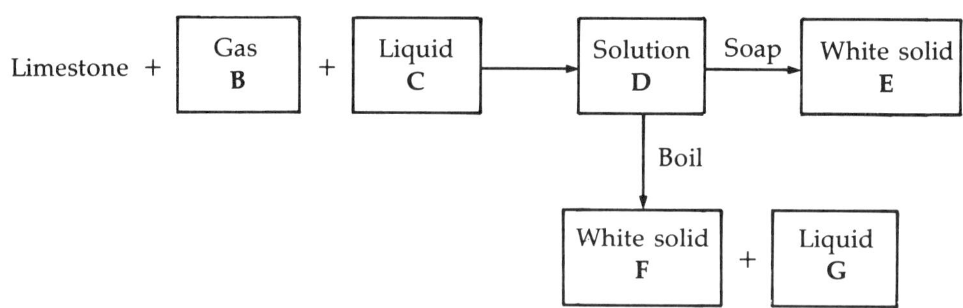

Questions Give the name *or* formula of

3 liquid **C**

4 solution **D**

5 solid **E** (name only)

6 solid **F**

7 liquid **G**.

Part 3 (Industrial)

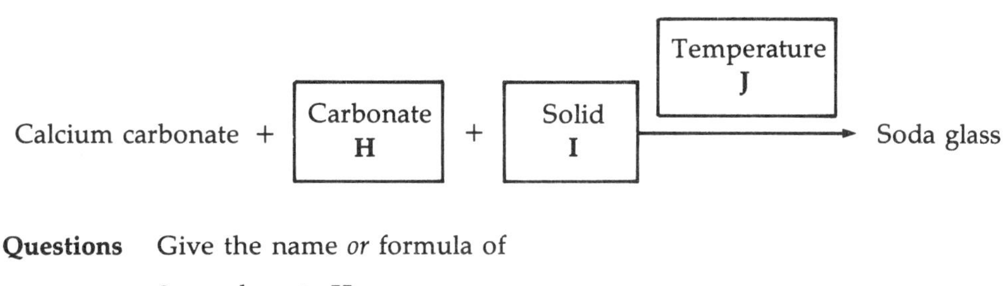

Questions Give the name *or* formula of

8 carbonate **H**

9 solid **I**.

10 What is temperature **J**?

REACTION SCHEME 8
Ammonia

Part 1 (Industrial process)

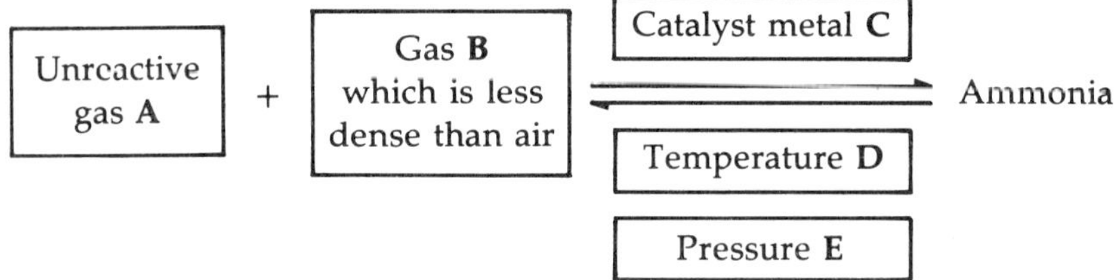

Questions

1 Give the name *or* formula of gas **A**.

2 Give the name *or* formula of gas **B**.

3 Give the name *or* formula of metal **C**.

4 What is temperature **D**?

5 What is pressure **E**?

Part 2

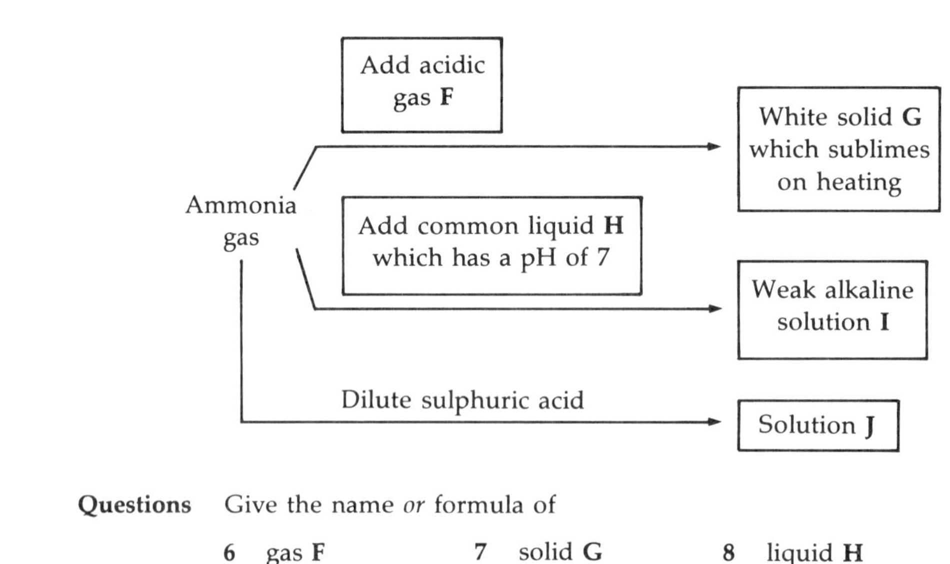

Questions Give the name *or* formula of

6 gas **F** 7 solid **G** 8 liquid **H**

9 solution **I** 10 solution **J**.

REACTION SCHEME 9
Effect of Heat on Nitrates

Part 1

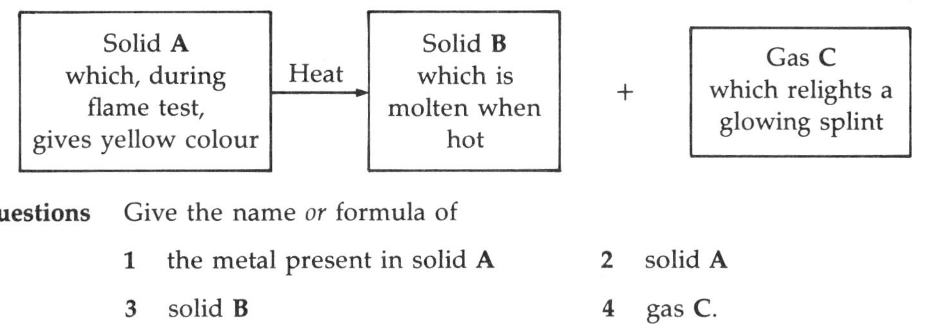

Questions Give the name *or* formula of

1 the metal present in solid **A**

2 solid **A**

3 solid **B**

4 gas **C**.

Part 2

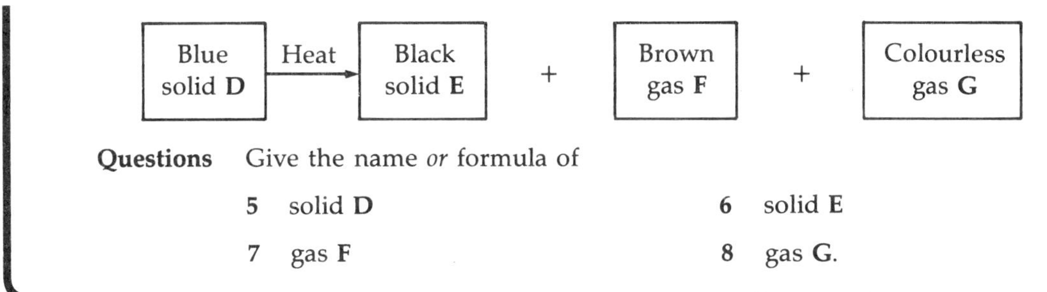

Questions Give the name *or* formula of

5 solid **D**

6 solid **E**

7 gas **F**

8 gas **G**.

Part 3

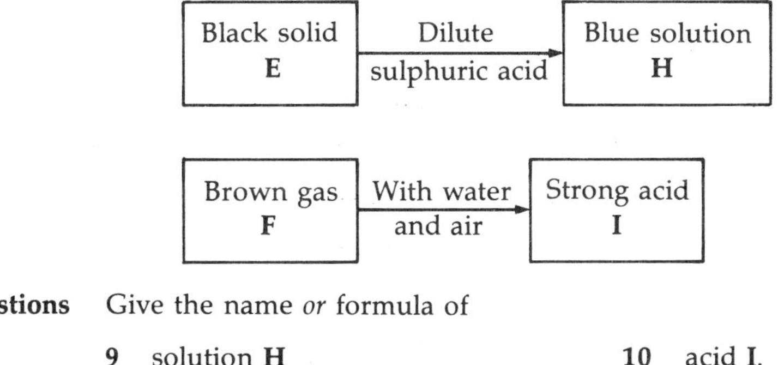

Questions Give the name *or* formula of

9 solution **H**

10 acid **I**.

REACTION SCHEME 10
Sulphur and Concentrated Sulphuric Acid

Part 1

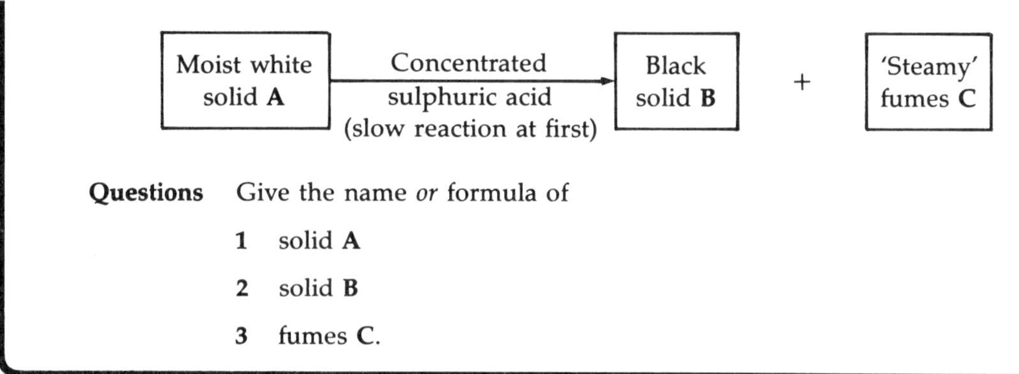

Questions Give the name *or* formula of

1 solid **A**

2 solid **B**

3 fumes **C**.

Part 2 (Industrial process)

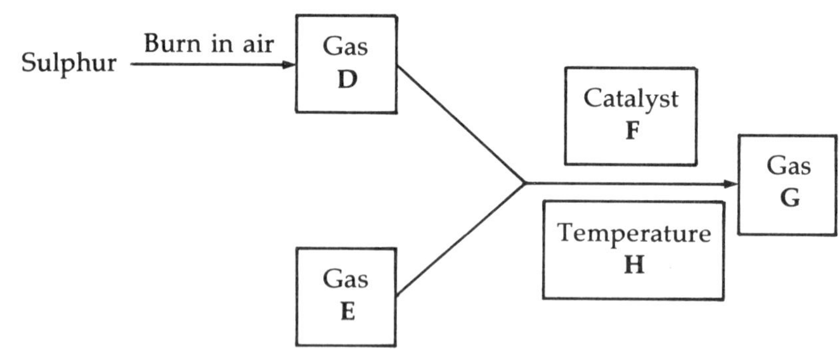

Questions

4 Give the name *or* formula of gas **D**.

5 Give the name *or* formula of gas **E**.

6 Give the name *or* formula of catalyst **F**.

7 Give the name *or* formula of gas **G**.

8 What is temperature **H**?

Part 3

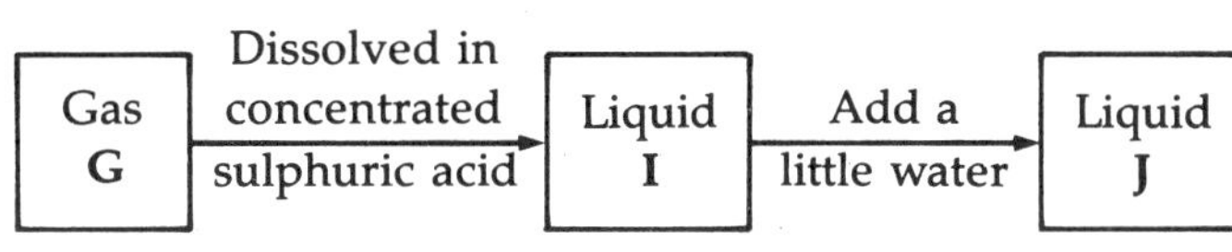

Questions Give the name *or* formula of

9 liquid **I**

10 liquid **J**.

REACTION SCHEME 11
Dilute Sulphuric Acid

Part 1

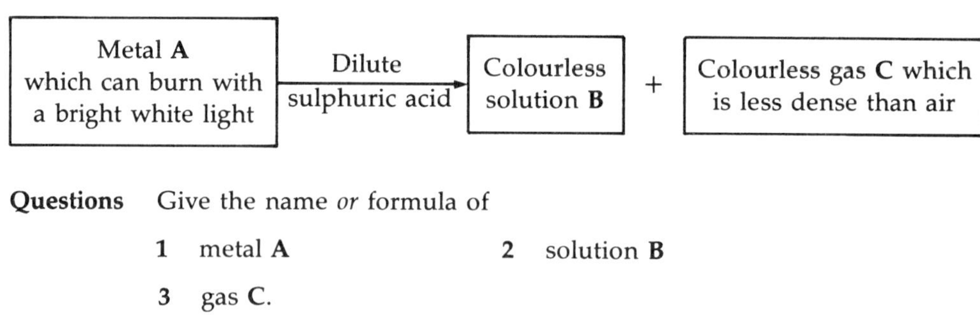

Questions Give the name *or* formula of

1 metal **A**

2 solution **B**

3 gas **C**.

Part 2

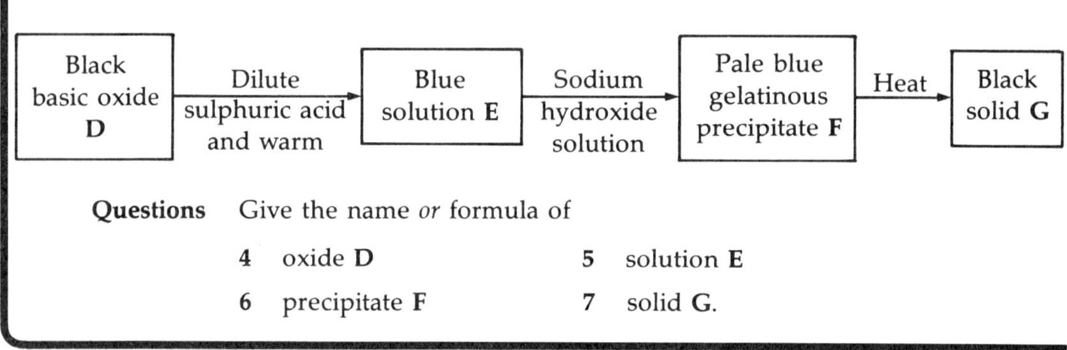

Questions Give the name *or* formula of

4 oxide **D**

5 solution **E**

6 precipitate **F**

7 solid **G**.

Part 3

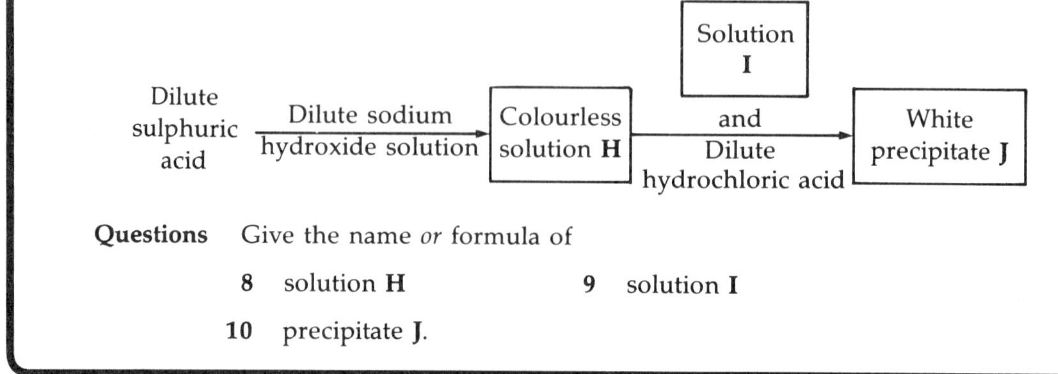

Questions Give the name *or* formula of

8 solution **H**

9 solution **I**

10 precipitate **J**.

REACTION SCHEME 12
Chlorine and the Halogens

Part 1

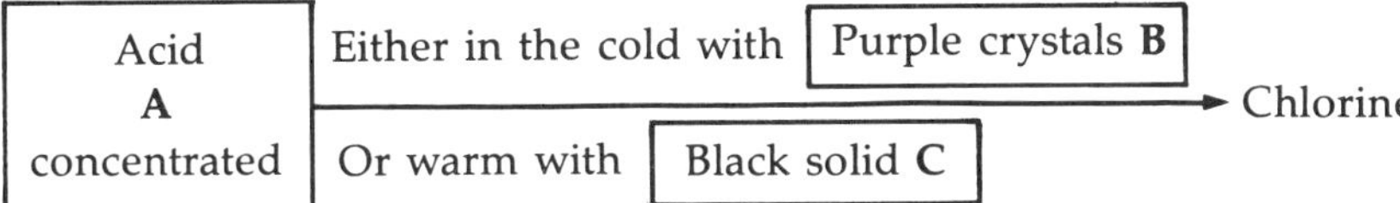

Questions Give the name *or* formula of

1 acid **A**

2 crystals **B**

3 solid **C**.

Part 2

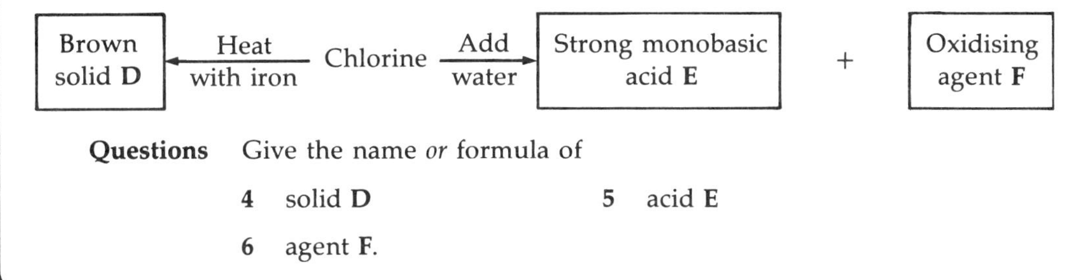

Questions Give the name *or* formula of

4 solid **D**

5 acid **E**

6 agent **F**.

Part 3

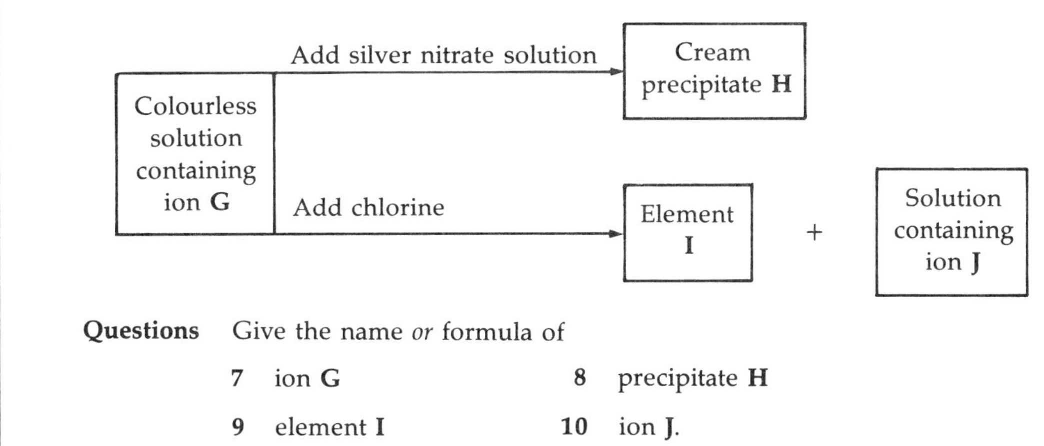

Questions Give the name *or* formula of

7 ion **G**

8 precipitate **H**

9 element **I**

10 ion **J**.

REACTION SCHEME 13
Hydrogen Chloride and Hydrochloric Acid

Part 1

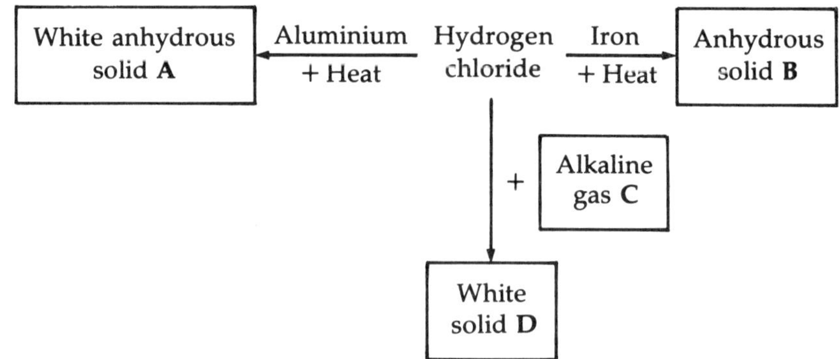

Questions Give the name *or* formula of

1 solid **A**

2 solid **B**

3 gas **C**

4 solid **D**.

Part 2

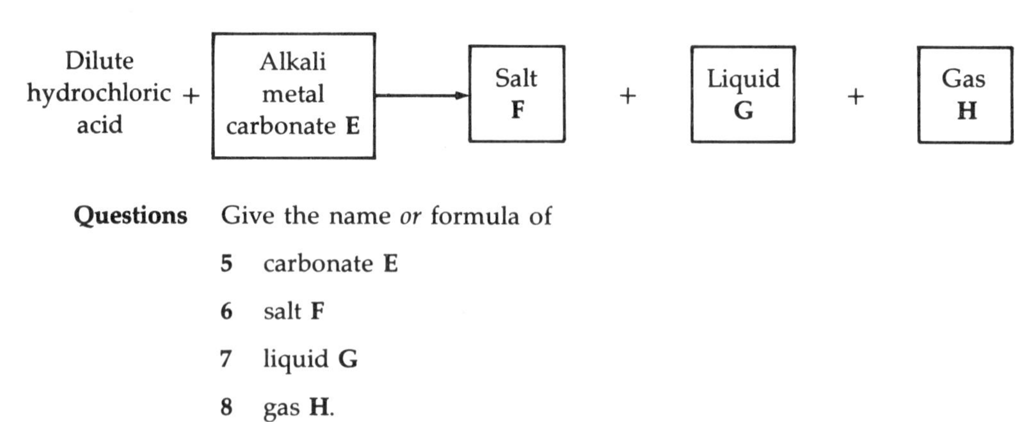

Questions Give the name *or* formula of

5 carbonate **E**

6 salt **F**

7 liquid **G**

8 gas **H**.

Part 3

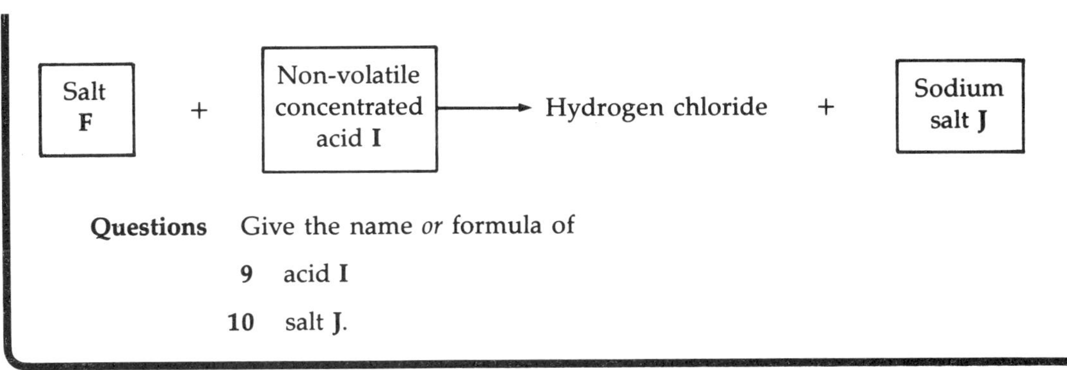

Questions Give the name *or* formula of

9 acid **I**

10 salt **J**.

REACTION SCHEME 14
Non-Metal Tests

Part 1

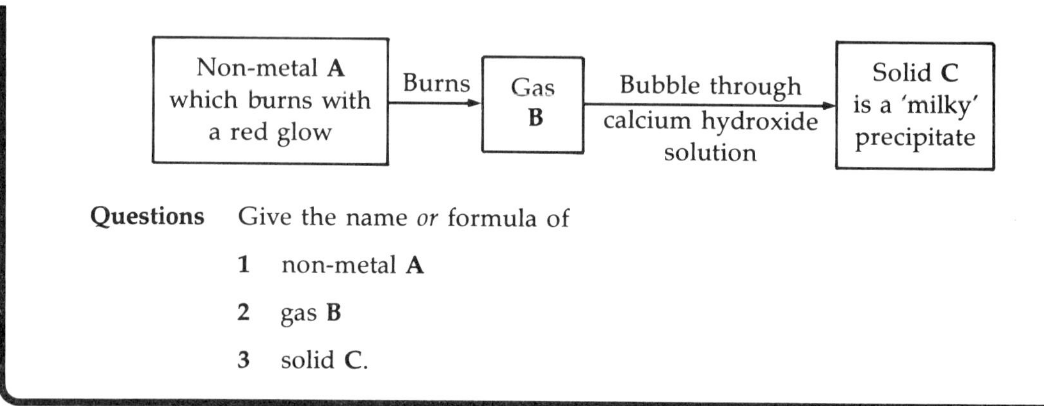

Questions Give the name *or* formula of

1 non-metal **A**

2 gas **B**

3 solid **C**.

Part 2

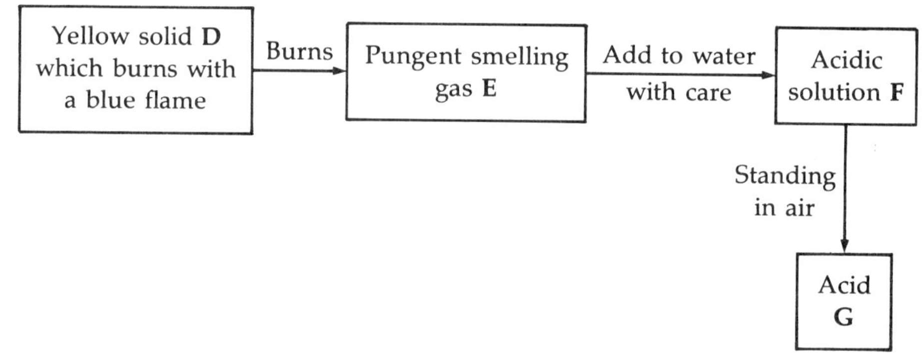

Questions Give the name *or* formula of

4 solid **D**

5 gas **E**

6 solution **F**

7 acid **G**.

Part 3

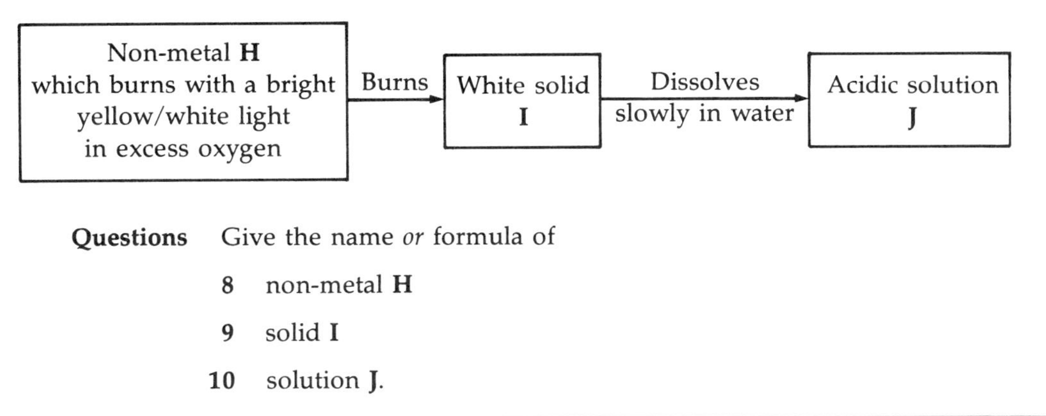

Questions Give the name *or* formula of

8 non-metal **H**

9 solid **I**

10 solution **J**.

REACTION SCHEME 15
Metal Flame Tests

Part 1

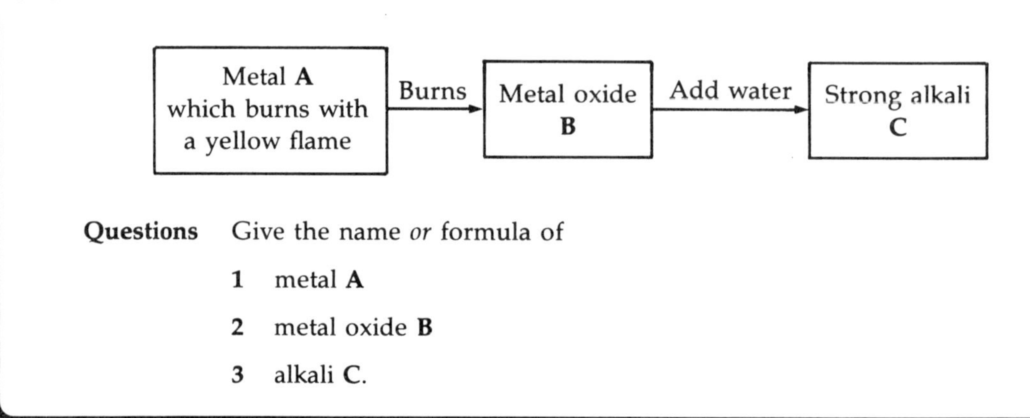

Questions Give the name *or* formula of

1 metal **A**

2 metal oxide **B**

3 alkali **C**.

Part 2

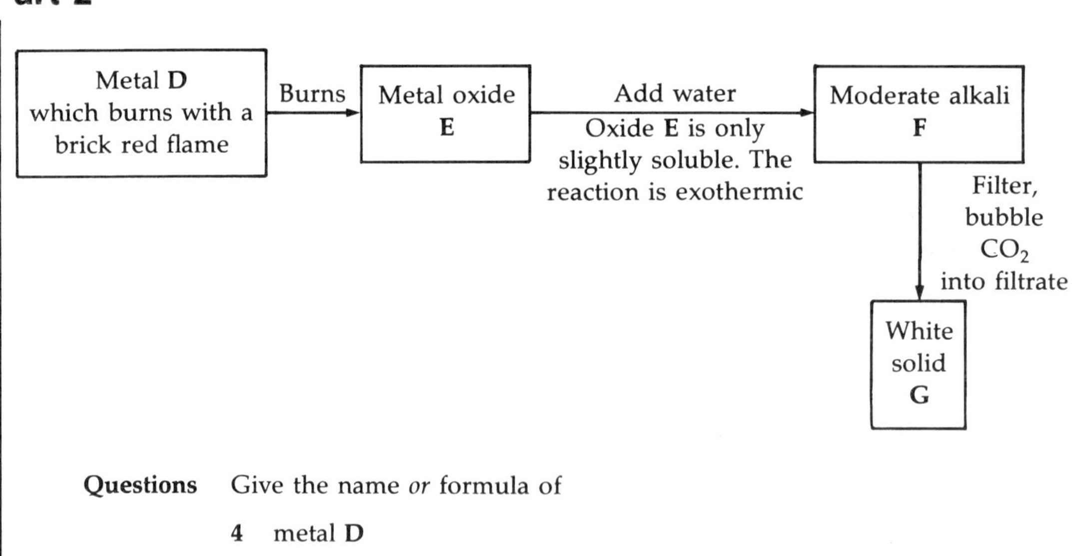

Questions Give the name *or* formula of

4 metal **D**

5 metal oxide **E**

6 alkali **F**

7 solid **G**.

Part 3

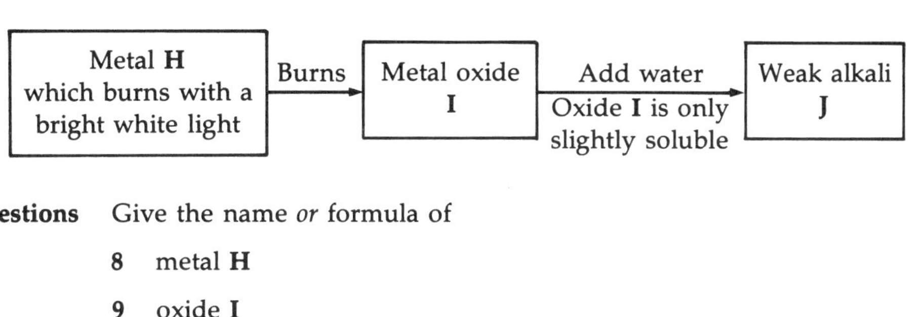

Questions Give the name *or* formula of

8 metal **H**

9 oxide **I**

10 alkali **J**.

REACTION SCHEME 16
Group I Metals

Part 1

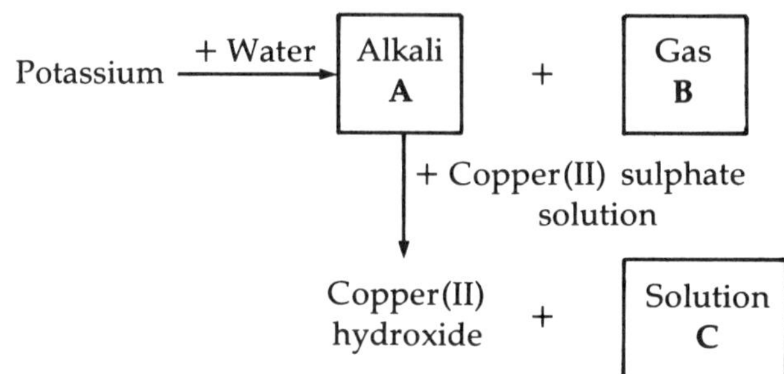

Questions Give the name *or* formula of

1 alkali **A** **2** gas **B**

3 solution **C**.

Part 2

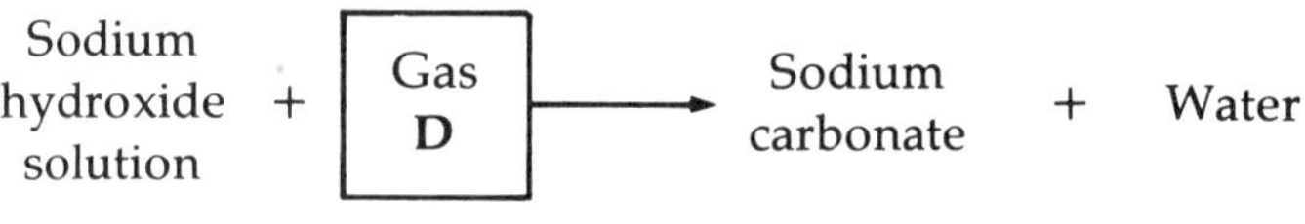

Question **4** Give the name *or* formula of gas **D**.

Part 3

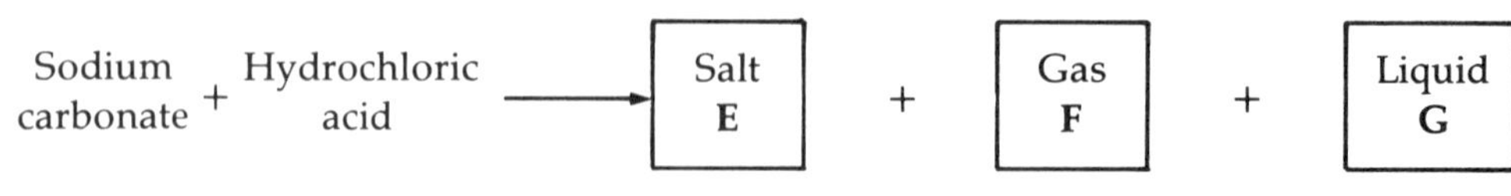

Questions Give the name *or* formula of

5 salt **E** **6** gas **F**

7 liquid **G**.

Part 4

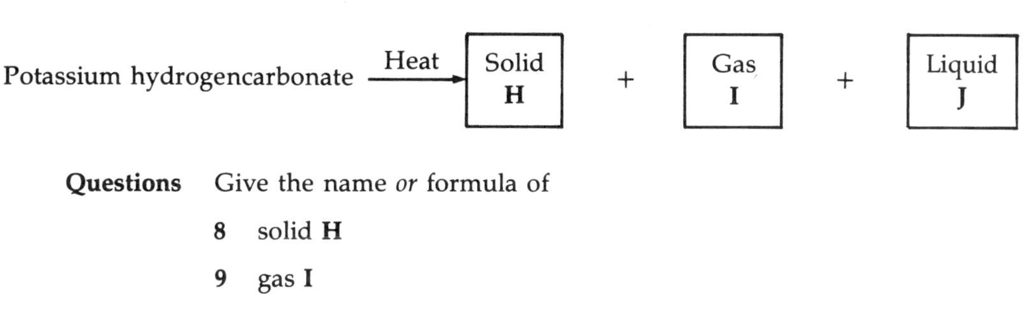

Questions Give the name *or* formula of

8 solid **H**

9 gas **I**

10 liquid **J**.

REACTION SCHEME 17
Magnesium

Part 1

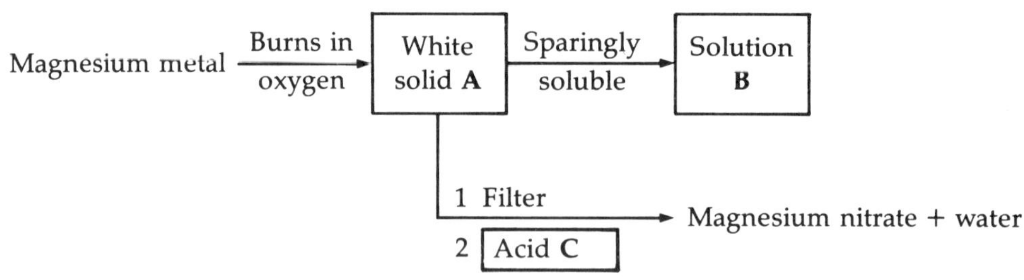

Questions Give the name *or* formula of

1 solid **A**

2 solution **B**

3 acid **C**.

Part 2

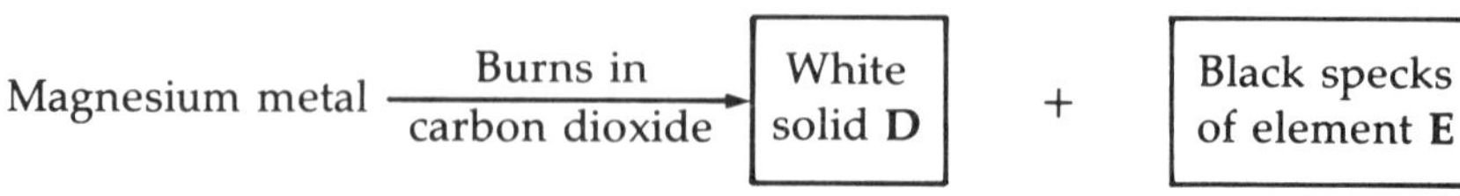

Questions Give the name *or* formula of

4 solid **D**

5 element **E**.

Part 3

Magnesium metal —Burns in nitrogen→ Black solid **F** —+ Water→ Alkaline solution **G** + Magnesium compound **H**

Questions Give the name *or* formula of

6 solid **F**

7 solution **G**

8 compound **H**.

Part 4

Magnesium metal + Steam ——→ Compound **I** + Gas **J**

Questions Give the name *or* formula of

9 compound **I**

10 gas **J**.

REACTION SCHEME 18
Iron

Part 1

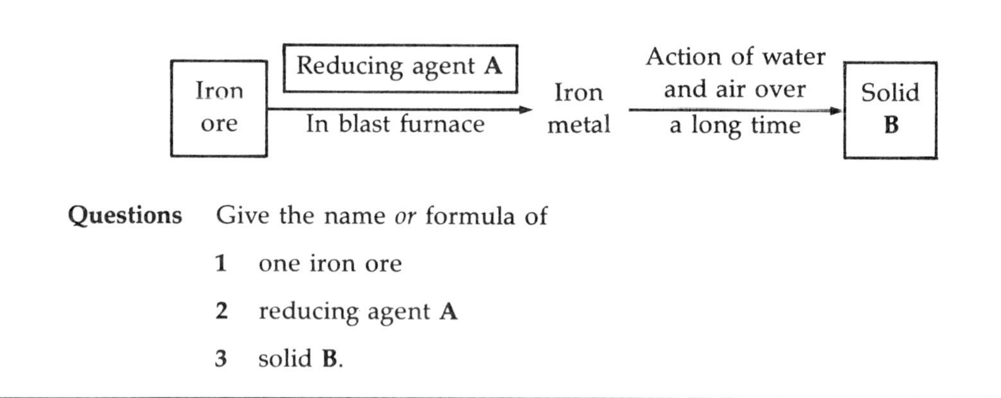

Questions Give the name *or* formula of

1 one iron ore

2 reducing agent **A**

3 solid **B**.

Part 2

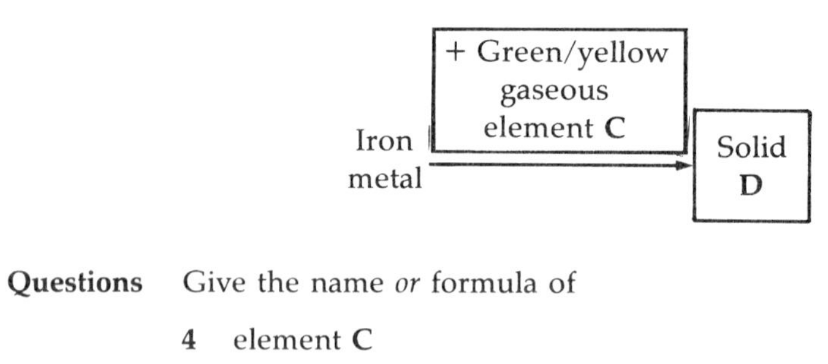

Questions Give the name *or* formula of

4 element **C**

5 solid **D**.

Part 3

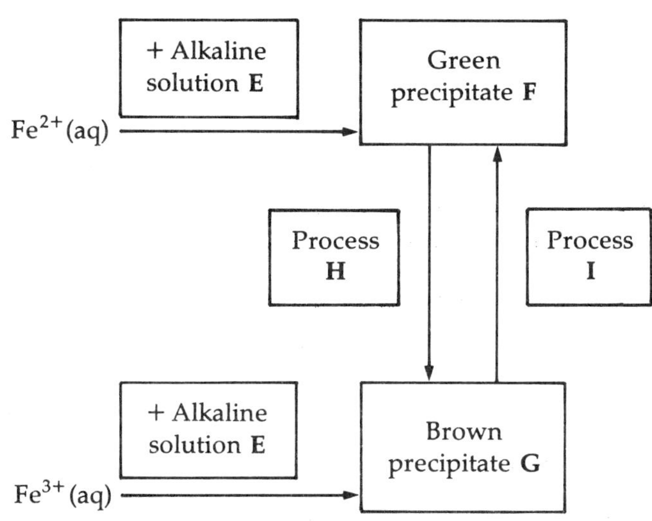

Questions Give the name *or* formula of

6 solution **E**

7 precipitate **F**

8 precipitate **G**.

Give the name of

9 process **H**

10 process **I**.

REACTION SCHEME 19
Copper

Part 1

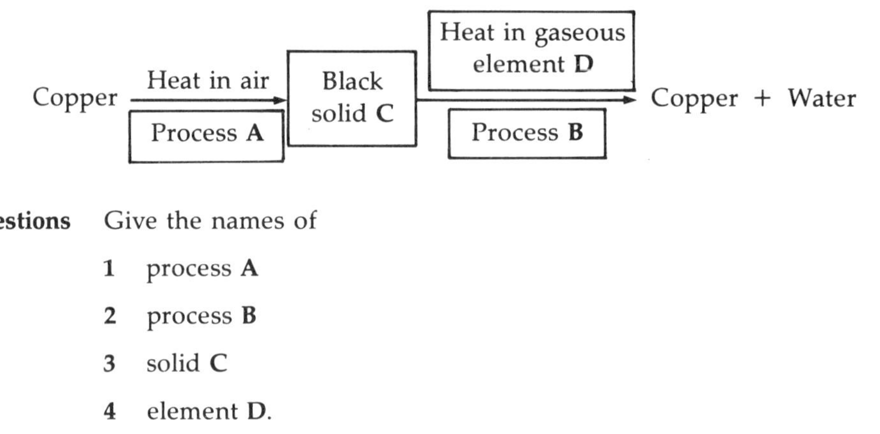

Questions Give the names of

1 process **A**

2 process **B**

3 solid **C**

4 element **D**.

Part 2

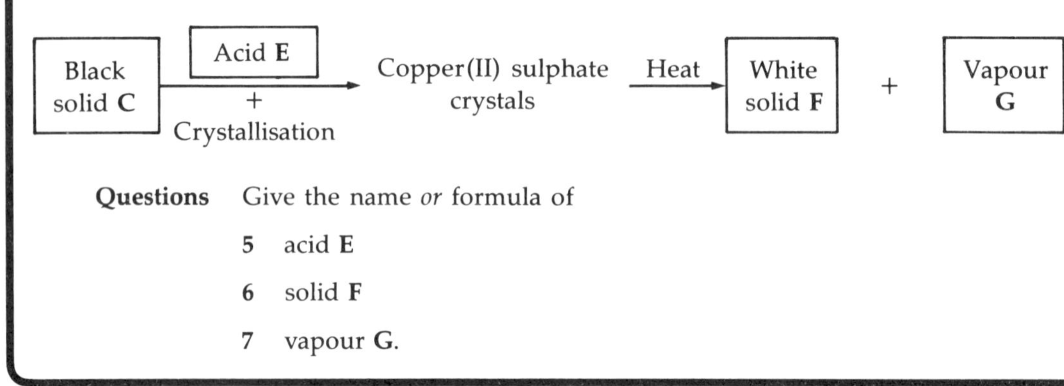

Questions Give the name *or* formula of

5 acid **E**

6 solid **F**

7 vapour **G**.

Part 3

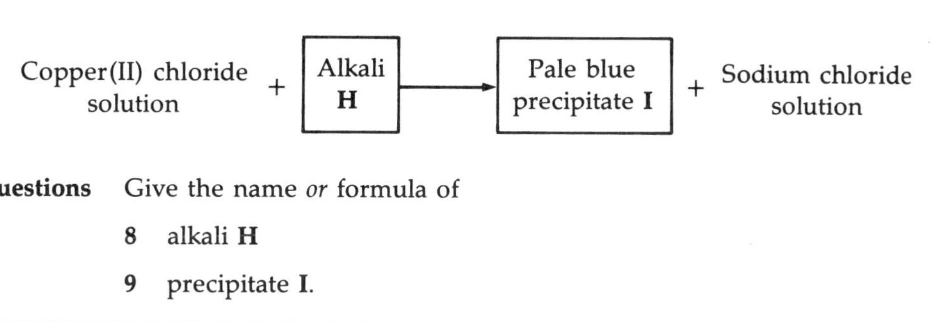

Questions Give the name *or* formula of

8 alkali **H**

9 precipitate **I**.

Part 4

Copper(II) nitrate solution + Magnesium metal ⟶ Copper metal + Solution **J**

Question **10** Give the name *or* formula of solution **J**.

REACTION SCHEME 20

Displacement reactions

Part 1

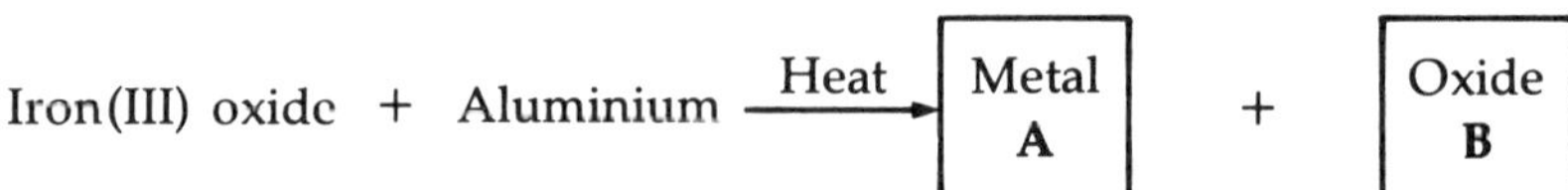

Questions

1 Give the name of metal **A**.

2 Give the name of oxide **B**.

Part 2

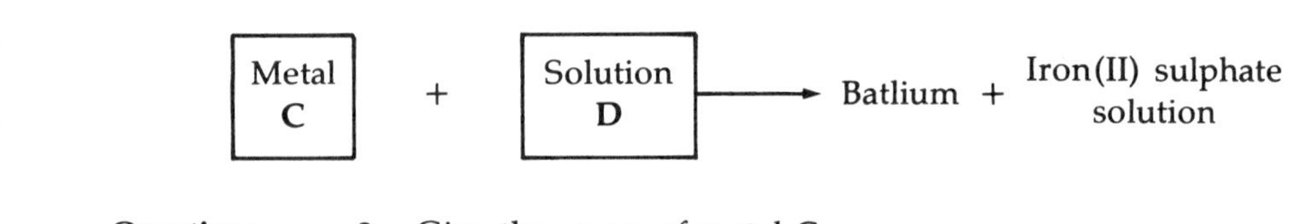

Questions

3 Give the name of metal **C**.

4 Give the name of solution **D**.

Part 3

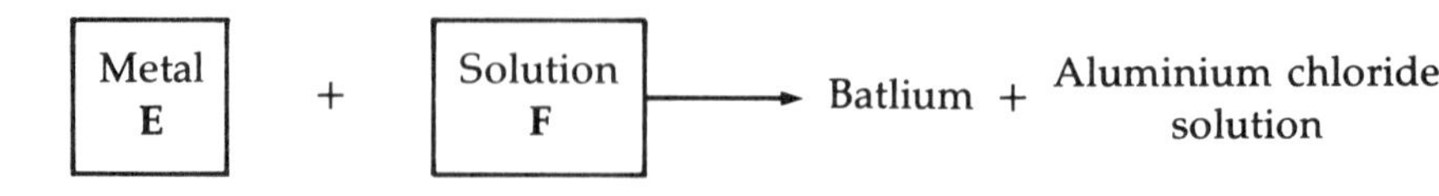

Questions

5 Give the name of metal **E**.

6 Give the name of solution **F**.

Part 4

Batlium + Silver oxide $\xrightarrow{\text{Heat}}$ [Metal **G**] + [Oxide **H**]

Questions

7 Give the name of metal **G**.

8 Give the name of oxide **H**.

Part 5

Questions

9 Place the four metals — silver, iron, batlium and aluminium — in their correct order of activity. Put the most reactive first.

10 Which of the above parts are 'thermit-type' reactions?

REACTION SCHEME 21
Hydrocarbons

Part 1

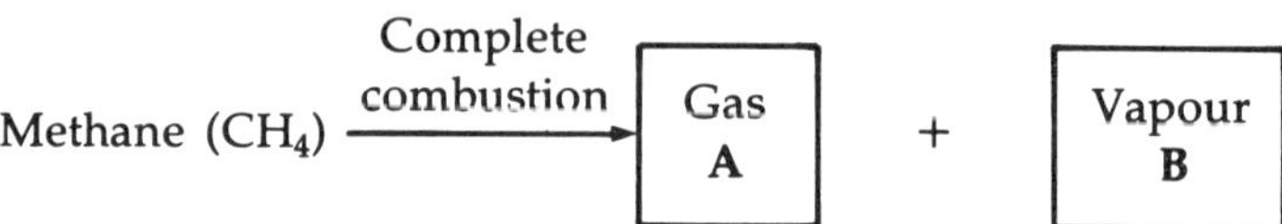

Questions Give the name *or* formula of

1 gas **A**

2 vapour **B**.

Part 2

Ethanol (C_2H_5OH) —(Concentrated sulphuric acid or Al_2O_3 + heat)→ Alkene **C** + Liquid **D** on cooling

Questions Give the name *or* formula of

3 alkene **C**

4 liquid **D**.

Part 3

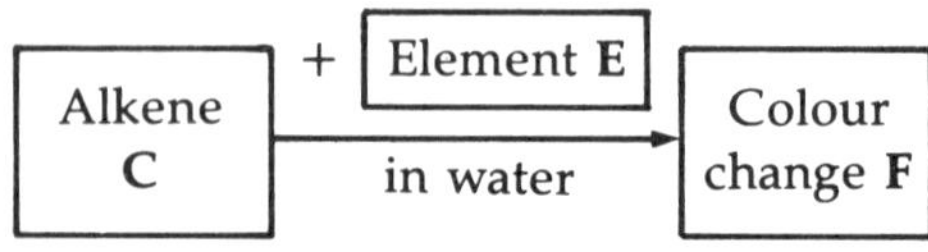

Questions **5** Give the name *or* formula of element **E**.

6 What is the colour change **F**?

Part 4

Process **H**

Propene (C_3H_6) + Hydrogen (H_2) ⟶ Hydrocarbon **G**

Questions **7** Give the name *or* formula of hydrocarbon **G**.

8 Name process **H**.

Part 5

Hexane (C_6H_{14}) —Hot vapour over broken pot / Process **I**⟶ Ethene (C_2H_4) + Alkane **J**

Questions **9** Name process **I**.

10 Give the name *or* formula of alkane **J**.

REACTION SCHEME 22
Ethanol

Part 1

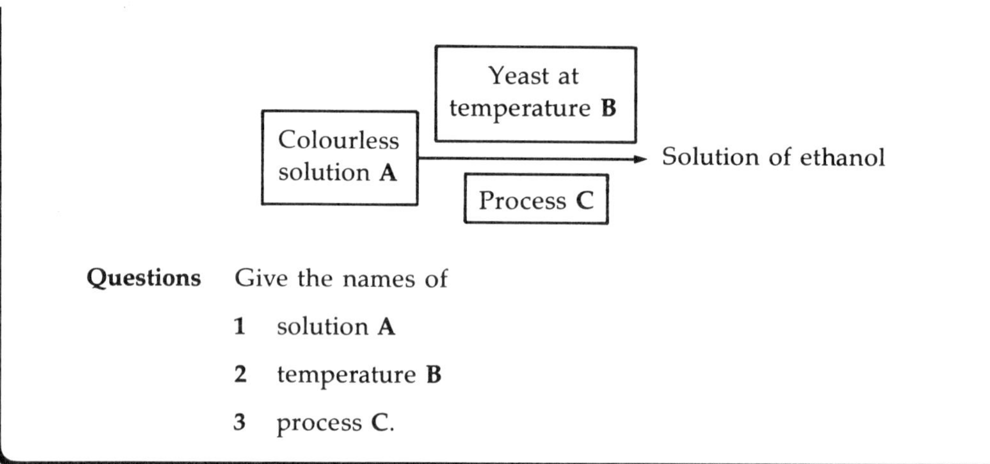

Questions Give the names of

1 solution **A**

2 temperature **B**

3 process **C**.

Part 2

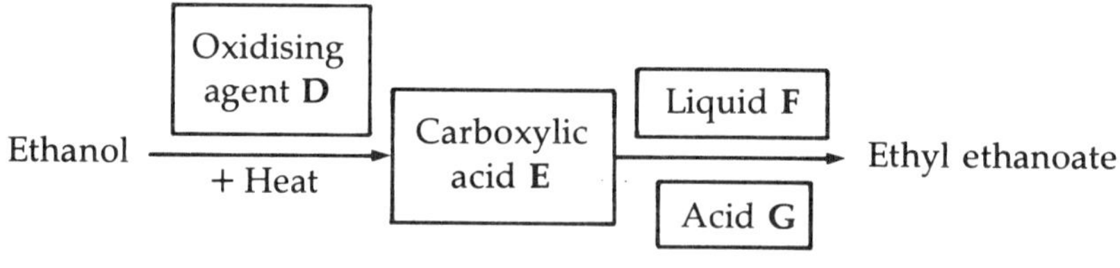

Questions Give the name *or* formula of

4 oxidising agent **D**

5 acid **E**

6 liquid **F**

7 acid **G**.

Part 3

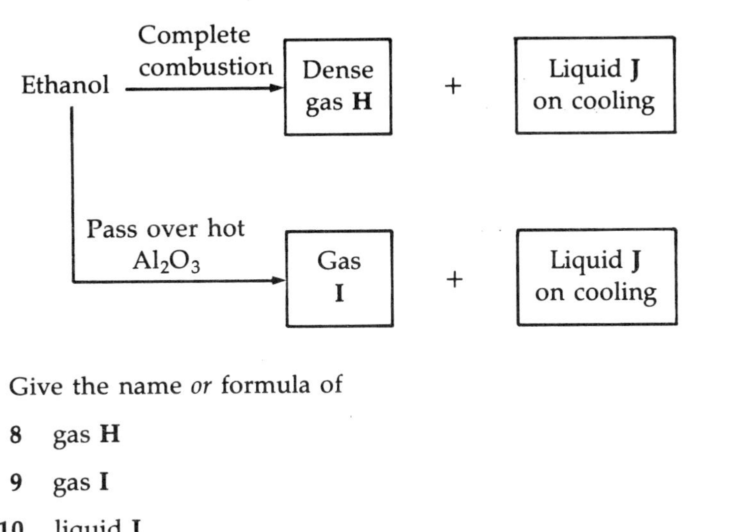

Questions Give the name *or* formula of

8 gas **H**

9 gas **I**

10 liquid **J**.

REACTION SCHEME 23
Electrolysis

Part 1

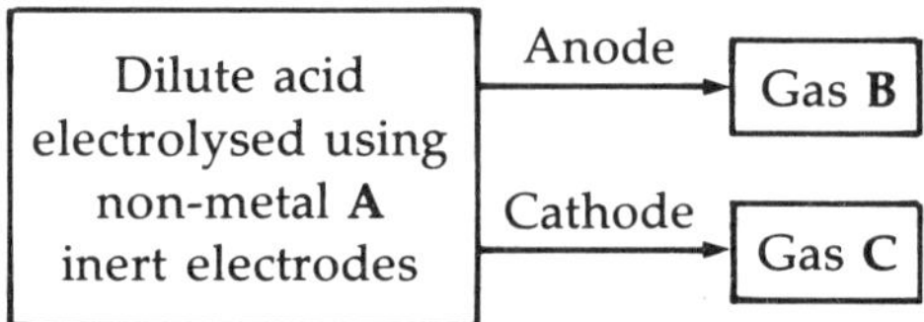

Questions

1 Give the name *or* formula of non-metal **A**.

2 Give the name *or* formula of gas **B**.

3 Give the name *or* formula of gas **C**.

4 What is the ratio of the volume of gas **C** to the volume of gas **B**?

Part 2

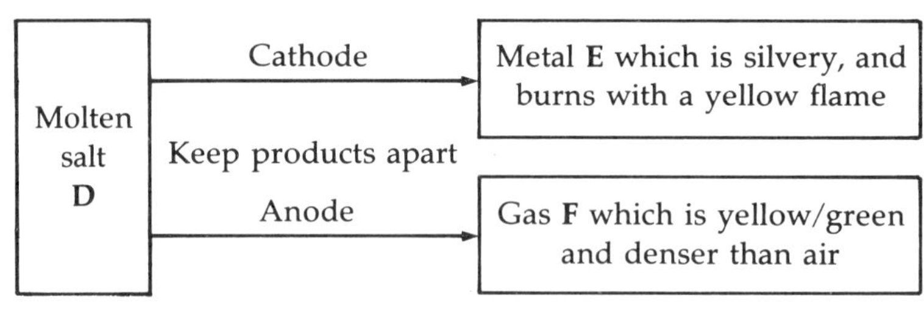

Questions Give the name *or* formula of

5 salt **D**

6 metal **E**

7 gas **F**.

Part 3

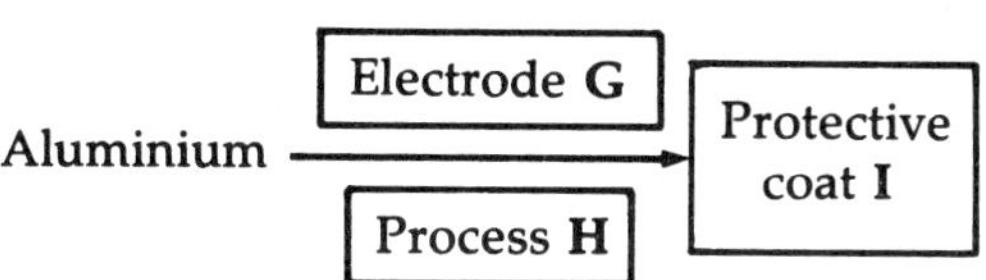

Questions

8 Name electrode **G**.

9 Name process **H**.

10 Name coat **I**.

REACTION SCHEME 24
Chemistry and the Environment

Part 1

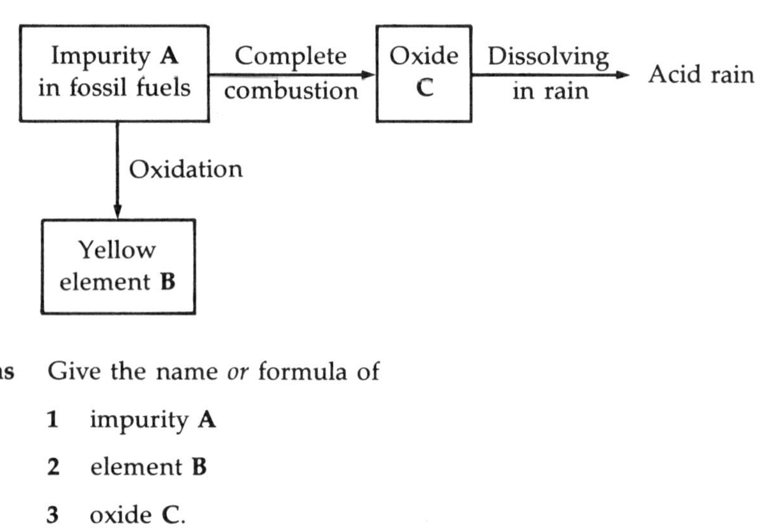

Questions Give the name *or* formula of

1 impurity **A**

2 element **B**

3 oxide **C**.

Part 2

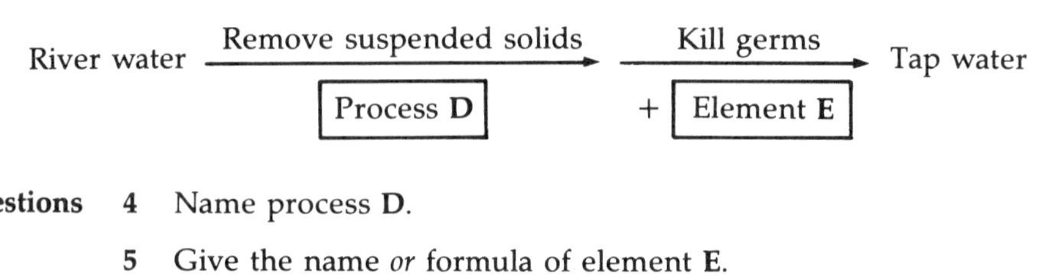

Questions **4** Name process **D**.

5 Give the name *or* formula of element **E**.

Part 3

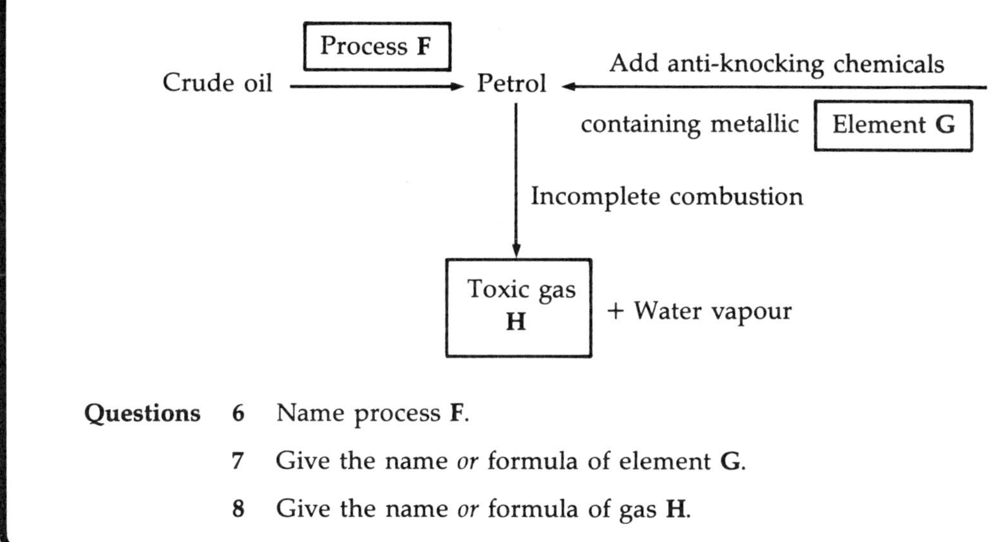

Questions

6 Name process **F**.

7 Give the name *or* formula of element **G**.

8 Give the name *or* formula of gas **H**.

Part 4

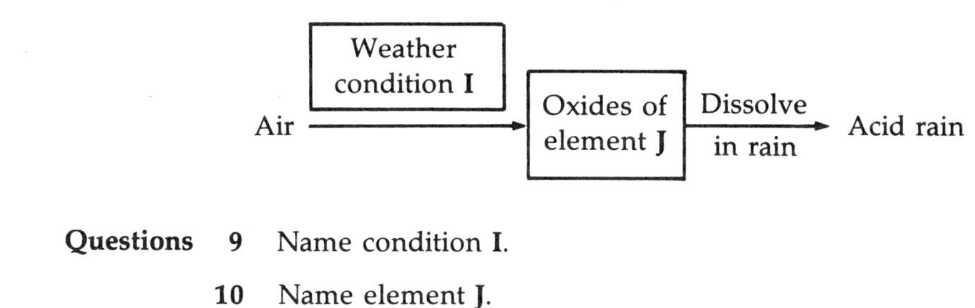

Questions

9 Name condition **I**.

10 Name element **J**.

REACTION SCHEME 25
Chemistry and Industry

Part 1

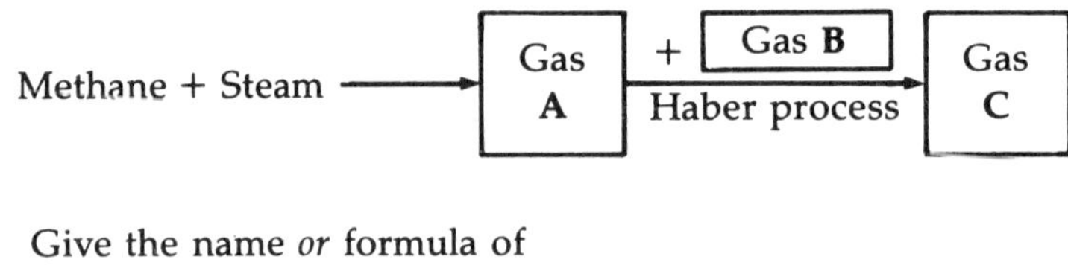

Questions Give the name *or* formula of

1 gas **A**

2 gas **B**

3 gas **C**.

Part 2

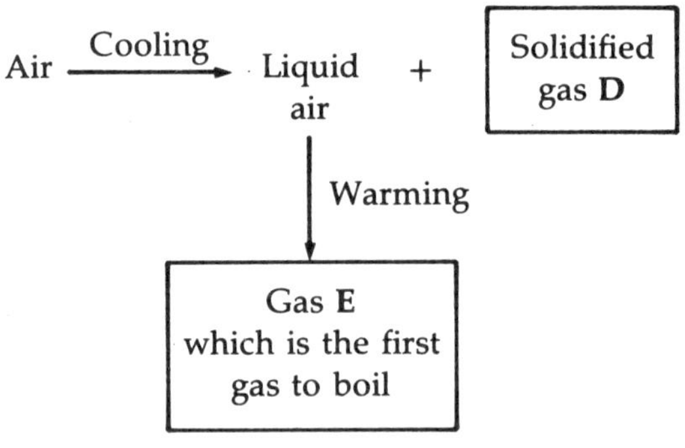

Oxygen boils at −183 °C
Nitrogen boils at −196 °C

Questions Give the name *or* formula of

4 gas **D**

5 gas **E**.

Part 3

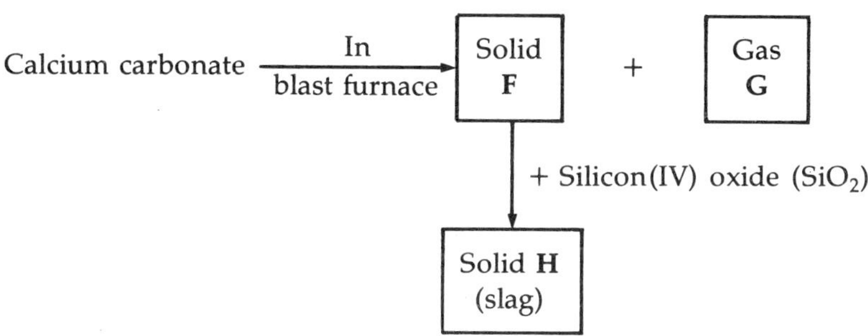

Questions Give the name *or* formula of

6 solid **F**

7 gas **G**

8 solid **H**.

Part 4

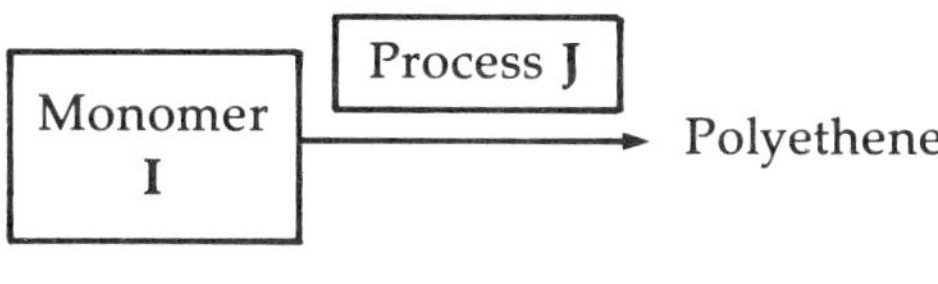

Questions Give the name of

9 monomer **I**

10 process **J**.